ALL ABOUT WEEDS

BY

EDWIN ROLLIN SPENCER

PH.D.

ILLUSTRATED BY

EMMA BERGDOLT

DOVER PUBLICATIONS, INC.

NEW YORK

This Dover edition, first published in 1974, is an unabridged republication of the expanded edition of the work as published by Charles Scribner's Sons, New York, in 1957 under the title *Just Weeds*.

International Standard Book Number: 0-486-23051-1
Library of Congress Catalog Card Number: 73-91485

Manufactured in the United States of America
Dover Publications, Inc.
180 Varick Street
New York, N.Y. 10014

90208

DEDICATED

TO

MY TWO SMARTWEEDS

MALCOLM AND JEAN SPENCER

This new edition is dedicated
to the memory of
EDWIN ROLLIN SPENCER
(1881–1964)
A beloved man and a dedicated naturalist,
scientist and educator

PREFACE

THERE are very few weed books that are of any use to the layman. Those who are prepared to write such books are usually so well versed in taxonomy that they write, if at all, for the botanists; or, if for the layman, they put their work in the form of bulletins. Some very fine bulletins have been published on such subjects as "Weeds Used in Medicine," "Thirty Poisonous Plants," "Unlawful and Other Weeds of Iowa," "Red Sorrel and Its Control," "White Snakeroot Poisoning," and many, many more. But to find how many men and women interested in weeds have read bulletins on weeds one has only to ask his next door neighbor and take the answer as indicative of the general practice. A survey of the homes in his city block, or of those of his farming community will convince any one that few indeed have seen any of these bulletins and that a much less number have read any of them.

One purpose of this book is to correct all of that. It is hoped that any one who becomes interested in a weed found in his garden or lawn, in wheat field or meadow, by the wayside or in a fence row may, by turning the pages of this book, learn the name of that weed, something interesting about it, and, if it is a bad one, how to get rid of it. It is hoped that an interest in many weeds may be developed in this way, and even a desire to learn to read the technical descriptions.

The author realizes that if the book succeeds as it should it will be largely due to the drawings by Miss Bergdolt. For three years, during the growing seasons, she patiently worked on the subjects suggested, and she redrew a subject every time the author pointed out the slightest defect. Sincerest thanks are due Miss Bergdolt. Thanks are also due Doctor J. M. Greenman, Taxonomist and

Curator of the Herbarium of the Missouri Botanical Garden, for his valuable help, and to Doctor George T. Moore, Director of the Missouri Botanical Garden, for the free use of the Garden's library and herbarium. The author wants also to thank his wife, Aileen Hunter Spencer, for her careful assistance in preparing the manuscript.

E. R. SPENCER

CONTENTS

II

WEEDS THAT ARE GRASSLIKE

(The order of arrangement of the plants is that found in Gray's *Manual of Botany*, but this list has been alphabetized for convenient reference)

III

WEEDS THAT ARE NOT GRASSLIKE

(The order of arrangement of the plants is that found in Gray's *Manual of Botany*, but this list has been alphabetized for convenient reference)

IV

WEED CONTROL

ILLUSTRATIONS

Page

I

THE REASONS FOR "JUST WEEDS"

WITH HABITAT AND SEASONAL INDEXES

I

THE REASONS FOR "JUST WEEDS"

OF ALL the forms of nature, unless it be insects, nothing is so sure to come into one's life as weeds. Most of this nation's population cannot step out of doors without being saluted by some weed of greater or less importance. To have a lawn or landscaped yard means to have trouble with weeds, and the farmer, the truck grower, the gardener, the orchardist, and even the greenhouse keeper must wage a continual war with these persistent plants. Any plant is a weed if it insists upon growing where the husbandman wants another plant to grow. It is a plant out of place in the eye of man; in the nice eye of nature it is very much in place. In the struggle for existence a bad weed is a prince. It has the traits of a Bonaparte or a Hitler. Give it an inch and it will take a mile, all because nature has endowed it with supervitality as well as with a few characters that make it useless to man and beast. That is the nature of the worst weeds. There are others with only a few weedy traits, and some with virtues that almost remove them from the weed list. The Bermuda grass is one of this sort. It makes a valuable lawn grass as well as an excellent hay and forage crop, but it can choke out valuable cultivated crops and by so doing it becomes a bad weed.

The principal purpose of this book is to teach. It is an attempt to make interesting to any and all readers a few of the most common forms of nature.

"To him who in the love of Nature, holds
Communion with her visible forms, she speaks
A various language."

There is a language of the weeds. They have their peculiarities and personalities, and this book was planned and written for the express purpose of helping the reader to see weeds as they are and not as just a mass of vegetation.

Most of the weeds treated here can be identified by the drawings alone. Miss Bergdolt, a rural schoolteacher of southern Illinois, entered the author's class of Local Flora in the summer of 1935, after the plan of this book had lain dormant for nearly twenty years. Instead of pressing specimens for her herbarium Miss Bergdolt drew pictures of the plants she studied. The author gave her a piece of Bristol board with the request that she draw thereon a picture of a dandelion. That was the first drawing for this book. It was also an answer to a long-felt prayer. All but three of the drawings contained herein are from life; two are from specimens found in the Herbarium of the Missouri Botanical Garden, and one is a redrawing from an illustration in Walter Conrad Muenscher's book, *Weeds,* by permission of the author and of the publishers, the Macmillan Company, of New York.

The aim of the drawings is to bring out the characters that untrained observers see when they look at plants. The form and venation of leaves, the general habit of plant growth, and the general appearance of flower or flower clusters are seen by every one; color differences and floral parts are seldom noticed except by botanists. The descriptive sketch accompanying each picture aims to call attention to the most prominent characters used in identification; to show, whenever possible, how the common as well as the scientific names are meant to be descriptive; to give some of the weedy ways of the pests, and to suggest methods of eradication and control. The sketches are followed by technical descriptions taken practically verbatim in all but one instance from *Gray's Manual of Botany.** The one exception is in the case of the Johnson grass, which comes from Hitchcock's *Manual of Grasses of the United States.*

Most readers of a book like this are interested in a very few weeds: the weeds of a lawn or yard, a weed in a vegetable or flower garden, a weed in a cornfield or truck patch, a few weeds by the wayside, and so on. It was knowing this fact that gave

New Manual of Botany, by Asa Gray; revised by B. L. Robinson and M. L. Fernald; seventh edition, 1908; copyrighted by American Book Company, New York. Quoted by permission of the publishers.

rise to the idea of arranging habitat indexes. There are but few bad weeds in any of the few general habitats. True, some of the worst weeds are found in nearly all places, but often the habitat so favors a weed that it becomes a nuisance there and nowhere else; as every one knows who has made the acquaintance of the chickweed, that pest of the lawn. Many weeds are like that. They are bad only when their environment favors them. A favorable environment becomes a favorite habitat.

Since there are not a great many weeds that stand out in any of the habitats, and since weeds may be divided into grass-like plants and those that are not grasslike, it is possible to shorten greatly the list to be scanned by the reader who has become interested in a weed of any given habitat. If he knows that the weed is grasslike; that is, has bladelike leaves such as those seen on plants like wheat, corn, oats, and blue grass, or leaves like those of the lilies and onions—if he knows the weed to have leaves such as these—he has only to look through the names of the grasslike weeds of the habitat index that most nearly describes the place where the weed flourishes. And even if he has to turn to every weed listed it will not be an arduous task. In nine cases in ten he will find the weed illustrated and described; in the tenth case it may be a local weed or it may be one of the very few widely distributed weeds for various reasons omitted.

What is true of the grasslike weeds is true of those that are not grasslike. The reader can easily find the plant even if he has to turn to every weed listed. The list is longer than that of the grasslike weeds but in many instances the names are descriptive enough to make the search easy.

This was the original plan of the book: the making of habitat indexes. Then came the desire to give to each weed treated enough of human interest to make a readable sketch; or at least enough of interesting facts to persuade the reader that a weed is worth knowing. That a knowledge of weeds is of value the author is not only willing but eager to declare. It is truly worth while to know any form of nature if for no other reason than to be able to commune with that particular form. But man must do more than commune with the forms of nature; he must use them. To fail to use a form of nature is to admit defeat at its hands, if it is an aggressive enemy. There is a reason, a utilitarian reason, for loving our enemies. If we are to fight weeds all our lives it matters not whether we know their names or personalities, but if we are to use them as they should be used we need to know and to love them. Nearly every farmer in the United States wastes valuable fertilizer every year when he permits a weed crop to go to seed on one of his cultivated fields, or when he mows that weed crop to keep it from going to seed. He hates weeds and so does not know their value. He will turn under a crop of sweet clover, but not a crop of weeds. He loves sweet clover, even though it is just a weed. He has been told how valuable this plant is because its roots are infected with a bacterium that is able to "fix" atmospheric nitrogen. It is true that the nitrogen used by this weed is taken from the air, but after the plant has used it to make its own protoplasm, and after that protoplasm has been destroyed by the soil bacteria, the results are exactly the same as when any other weed is plowed into the soil. It is the decay of the protoplasmic material that makes available the nitrates that all plants must have in order to synthesize their own protoplasm. The nitrates made from the protoplasmic contents of a ragweed, or any other weed, are identical with the nitrates derived from the protoplasmic contents of sweet clover, or any other clover, or any other legume. The only difference to be pointed out is that legumes have in their protoplasm nitrogen that was once in the air, while non-leguminous plants must get their nitrogen as nitrates from the soil; but for some reason or other the leguminous weeds seem to be able to get their nitrates where the non-leguminous crop plants fail to get them, or at least where they fail to get them in sufficient quantities to be of much worth to the crop plants. That is weed nature. That is what we mean when we say that a weed has supervitality. Well then, when weeds are plowed under they decay just as sweet clover decays (if the ground is as sweet as it has to be to grow sweet clover) and the nitrogen they used in the making of their leaves and stems is made available to the crop plants that follow the weeds.

If there were no other reasons for knowing weeds their soil-building potential would suffice. But there is something of far more importance than soil building if we consider human health the most important thing in the world. Weeds were the mother of medicine. It is surprising how many weeds are still found listed in the pharmacopœias of the world. Even the dandelion is among the medicinal plants, and there is catnip, burdock, mustard, horehound, Jimson weed, and a great many more that are not so well known. Some of them have been dropped from the pharmacopœias but their extracts and tinctures are still to be found in the *U. S. Dispensatory,* a book listing all of the available medicinal compounds. Materia medica had its beginning among the weeds. It was early discovered that plants possessed healing properties. Drowning persons grasp at straws; sick people pull weeds, and from savage to sage relief has been obtained thereby. Civilized man has never been able to do without his vegetable compounds, and most of the contents of these compounds are derived from just weeds.

The supposed potency of a weed is often reflected in its name. For instance, the botanical name of the yarrow, a common meadow weed, is *Achillea millefolium,* which means the thousand-leaved plant used by Achilles. It is said that the leaves of this plant will stanch blood and that Achilles used its leaves on the wounds of his soldiers. Evidently he could not find any yarrow when that arrow from the bow of Paris struck him in the heel. Anyway, the name gives some idea of the plant's mythical healing powers. One of its English names is Bloodwort, which simply means blood plant, and which, of course, refers to the astringent nature of the juice of the plant. The crushed leaves are said to be effective in stopping nosebleed.

Many of the weeds treated in this book are among the medicinal plants. It may be disconcerting to him who relies on patent medicines to learn that much of their effectiveness is derived from the juices and extracts of just weeds, but such is the case.

The control of weeds is the principal concern of most people who have anything to do with the pests. How can I get rid of chickweed, the dandelions, creeping Jenny, Canada thistle, and the rest?

There are ways to fight and control weeds, and the best-known methods are given in the sketches of the most pestiferous weeds treated in this book. There are no easy ways and no cheap ways to fight weeds. Several of them have been outlawed in several of the States, and if we were law-abiding citizens it might be possible for us to eradicate, totally, some of our worst weed enemies and thus relieve ourselves, for all time, of their inroads and robberies. Bad weeds take a terrific toll every year from our agricultural interests. We complain of taxes but say nothing when Johnson grass takes a cotton, a corn, or a potato field, or when wild garlic causes a dockage in the milk or wheat prices, even though that may be many times higher than the highest of tax levies. We work and fail to produce a decent lawn all because of that brazen hussy, the creeping Jenny, a weed that is a federal outlaw. We go right on paying her bills and caring for her until not a home in our little city has a decent lawn.

WEEDS OF THE LAWN AND YARD

I. WEEDS THAT ARE GRASSLIKE

II. WEEDS THAT ARE NOT GRASSLIKE

WEEDS OF THE GARDEN AND TRUCK PATCH

WEEDS OF THE MEADOW AND PASTURE LANDS

I. WEEDS THAT ARE GRASSLIKE

II. WEEDS THAT ARE NOT GRASSLIKE

WEEDS OF THE CORN AND COTTON FIELDS

I. WEEDS THAT ARE GRASSLIKE

II. WEEDS THAT ARE NOT GRASSLIKE

WEEDS OF WINTER WHEAT AND CLOVER FIELDS

I. WEEDS THAT ARE GRASSLIKE

II. WEEDS THAT ARE NOT GRASSLIKE

Weeds (*Not Grasslike*) of Winter Wheat and Clover Fields (*continued*)

WEEDS OF THE FARM LOTS

I. WEEDS THAT ARE GRASSLIKE

II. WEEDS THAT ARE NOT GRASSLIKE

WORST WEEDS OF WAYSIDE AND WASTE PLACES

I. WEEDS THAT ARE GRASSLIKE

Worst Weeds of Wayside and Waste Places (*continued*)

II. WEEDS THAT ARE NOT GRASSLIKE

WEEDS OF MOIST AND WET PLACES

I. WEEDS THAT ARE GRASSLIKE

II. WEEDS THAT ARE NOT GRASSLIKE

WEEDS OF SPRING TIME

I. WEEDS THAT ARE GRASSLIKE

(The weeds that bloom and seed in the spring are not a great many. Most of the weedy grasses are then to be seen as green blades only. The grasses here given are those that start early enough to attract attention.)

WEEDS OF SUMMER

(*Weeds that bloom and seed in summer.*)

Weeds (*Grasslike*) of Summer (*continued*)

II. WEEDS THAT ARE NOT GRASSLIKE

(Summer is the weed season, and nearly all of our weeds, even those that bloom in the spring, bloom again in summer. Only the most conspicuous weeds of the season are listed here.)

WEEDS OF AUTUMN

(*Weeds that bloom or seed in the fall.*)

I. WEEDS THAT ARE GRASSLIKE

II. WEEDS THAT ARE NOT GRASSLIKE

WEEDS OF WINTER

(*Most of the summer and autumn weeds that have woody stems may be seen as dead stalks throughout the winter, but the weeds of the winter here given are seen as green plants.*)

I. WEEDS THAT ARE GRASSLIKE

II. WEEDS THAT ARE NOT GRASSLIKE

II

WEEDS THAT ARE GRASSLIKE

FIG. 1. Broom Sedge, a grass and not a sedge

BROOM-SEDGE

[*Andropogon virginicus* L.]

BROOM-SEDGE, or Broom-sage as the Southerner calls it, is neither a sedge nor a sage. It is a grass and it is very conspicuous in the fall of the year in all of the eastern and southern States. It is conspicuous because of the reddish color of its straight-stemmed bunches which wave like red flags in the late autumn and winter winds. By late fall the bearded seeds which give the name *Andropogon* to the genus are conspicuous also.

Broom-sedge is a weed, but a weed with a few virtues. It is called broom-sedge because early housewives made brooms of it. It makes a very good pasture if kept from reaching the seeding stage, and it can even be used for hay if cut before the stems become woody, which means before they are fully developed. It is a perennial and so persists for some time in a well-grazed pasture even if it is not permitted to set seed, and in pastures where the soil is too sour to produce other forage crops it often serves the stockman fairly acceptably.

The grass is sometimes called Beard grass—the very best name it has—and Sedge-grass. It has been called sedge, no doubt, because of its straight stems. Sedge stems are straight but they are not jointed as are grass stems, and sedges do not bear the type of leaves that the grasses do. The name, sedge, is merely a mistake in identity, but when it is called Broom-sage, that is a corruption.

The plant is not easily distinguished in its early stages, but any one who cares to know the weed can gratify his ambition when it reaches maturity. Then the red bunches, not brilliant red, of course, but red enough to contrast with all surrounding vegetation, loom up, and these bunches, three or more feet in height, actually wave like red flags in the wind. Some one has said that when the Broom-sage waves its red flag in the face of the farmer he has every reason to copy the behavior of his bull under like circumstances. It is time for that farmer to begin to fight if he wants to grow anything but Broom-sedge on his farm. He must

23

now spread something different from what he has been spreading. He must spread lime, limestone, sludge, marl, anything that will sweeten his soil and make it tolerant of his crop plants.

The name, *Andropogon,* means bearded like an old man. It refers to the white beardlike covering of the seeds. *Virginicus,* the specific name, is in honor of the State of Virginia. Linnæus evidently thought the weed originated there, or that more of the grass was to be found there than anywhere else.

TECHNICAL DESCRIPTION

Andropogon (Royen) L. Spikelets in pairs (one sessile and perfect, the other pediceled, sterile, often rudimentary) at each joint of the articulate rhachis; glumes of fertile spikelet subequal, indurated, the first dorsally flattened, with a strong nerve near each margin, the mid-nerve faint; second glume keeled above; first lemma empty, hyaline; fertile lemma membranaceous or hyaline, awned; palea hyaline, sometimes obsolete.—Tall tufted perennials; spikes lateral and terminal, the rhachis and usually the pedicels long-villous with silky hairs (whence the name composed of two Greek words meaning *man* and *beard*).

Andropogon virginicus L. *Culms rather slender,* 5–12 dm. high, *sparingly branched above;* sheaths smooth or somewhat hirsute on the margin; blades usually hirsute above near the base; *spathes smooth;* racemes 2 or 3, slender; hairs long and silky.—Open ground, Massachusetts to Illinois, Florida, and Texas.

JOHNSON GRASS
[*Sorghum halepense* Pers.]

WHERE Johnson grass has become a pest no one has to be told what it is, but where it is known only through its reputation those who live in fear of it ought to inform themselves, so that they may not be guilty of destroying harmless grasses in its name.

First of all, no grass is Johnson grass that is not more than three feet tall when it is fully grown. A tall, rank grass with a panicle ("head") that suggests a loose "head" of the sorghum plant is likely to be Johnson grass, and if such a grass comes into

FIG. 2. The Johnson Grass and one of its rootstalks

the community it should be watched. It is said to be a very fine grass for hay, and for this reason many Southern farmers sowed it when it was first introduced. One man says that if a farmer has no intention of raising anything but hay on his place, and at the same time has no regard for his neighbors' rights, he might be advised to sow Johnson grass, but otherwise he should never so much as think of doing it. It is not only almost impossible to eradicate Johnson grass when it is once established, it is next to impossible to raise anything else where it is. Cotton fields, potato fields, and cornfields are often taken by the weed. It can be eradicated, but to do it takes a regular program of eradication carried out to the letter. First cut the grass just before it blooms and make hay of it or stack it where it will be eaten by stock. As soon as the hay is off break the ground by plowing deep. Harrow out all the rootstalks possible. Do this two or three times during the growing season, and just before frost plow the field again and leave as many of the rootstalks as possible on top of the soil. Let the ground fallow all winter and plant the field to some cultivated crop the next spring. This procedure should completely eradicate the Johnson grass.

The plant has another weedy character that is peculiar to the sorghums. He who decides to use Johnson grass as a hay or pasture should remember that when frosted the leaves develop a poison that is fatal to cattle. This same poison (Prussic acid) often develops in the second crop of Johnson grass, especially when the season is hot and dry.

Johnson grass has a tall coarse stalk, wide coarse leaves that have a conspicuous midvein, and the panicle or "head" is coarse and has a reddish color that is just a little different from that of any other grass. It is almost a wine color. The picture gives the shape and appearance of the head. The plant has a rootstalk on which it depends for much of its propagation. The rootstalks break up when the ground is cultivated, and every piece, even if no more than an inch in length, will produce one or more new stalks of the weed. If the reader thinks he has a clump of Johnson grass to deal with he has only to dig down among its roots to prove his conclusions. If the big rootstalks are there it is Johnson grass; if they are not, it is likely Sudan. Another identifying mark is the red blotches on the leaves caused by bacteria. Sorghums are all more or less susceptible to this bacterial disease, but Johnson grass is one of the most susceptible, and one seldom sees it, especially after it is fully grown, without these red blotches.

Johnson grass has several other common names. It is called Evergreen millet, Egyptian millet, Morocco millet, Maiden cane, Syrian grass, Cuba grass, St. Mary's grass, Mean's grass, and Millet grass. The botanical name, *Sorghum halepense,* means the sorghum from Halepa, the section of Syria from which the grass is supposed to have come. *Sorghum* was the Latin name for the cane used in making syrup and Johnson grass belongs to the sorghum genus.

TECHNICAL DESCRIPTION

Sorghum Moench. Spikelets in pairs, one sessile and fertile, the other pedicellate, sterile but well developed, usually staminate, the terminal sessile spikelet with two pedicellate spikelets. Tall or moderately tall annuals or perennials, with flat blades and terminal panicles of 1- to 5-jointed tardily disarticulating racemes. Name from *Sorgho,* the Italian name of the plant.

Sorghum halepense (L.) Pers. Culms 50 to 150 cm. tall, from extensively creeping scaly rhizomes; blades mostly less than 2 cm. wide; panicle open, 15 to 50 cm. long; sessile spikelet 4.5 to 5.5 mm. long, ovate, appressed-silky, the readily deciduous awn 1 to 1.5 cm. long, geniculate, twisted below; pedicellate spikelet 5 to 7 mm. long, lanceolate. Open ground, fields, and waste places, Massachusetts to Iowa and Kansas, south to Florida and Texas, west to southern California; native of the Mediterranean region, found in the tropical and warmer regions of both hemispheres. Cultivated for forage; on account of the difficulty of eradication it becomes a troublesome weed.

(The above descriptions taken from Hitchcock, *Manual of the Grasses of the United States,* published by the United States Department of Agriculture.)

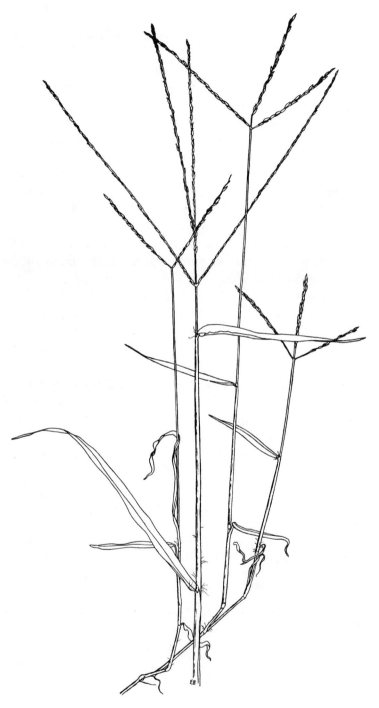

FIG. 3. The Crab Grass, a lawn
and garden pest

CRAB GRASS

[Digitaria sanguinalis Scop.]

NEARLY every one who has had anything to do with a garden knows what Crab grass is. If he does not know it as Crab grass he probably knows it as Finger grass, Polish millet, Crowfoot grass, or Pigeon grass.

There are two species of Crab grass, but they are very much alike except in size, and since the big one is the worse one it is the subject of this sketch.

The grass evidently got its name because its branching head reminded some one of a crab with its many legs. Finger grass is even more descriptive, and that is what the scientific name *Digitaria* means. *Digitaria sanguinalis* means the sanguine finger grass, and if sanguine means ardent and filled with enthusiasm, that name fits the plant perfectly. Clean up a garden and leave it for a week or two and the Crab grass will demonstrate its botanical name. Its sanguine bunches will be spreading their fingers all over the place when the sojourn ends.

Crab grass is one of the fastest-growing weeds. It also possesses another weedy character that few weeds have so fully perfected. It can grow and re-establish itself if only one tiny root is left in the ground. Many times the hoe fails to get every root, but from all appearances the weed has been hoed out. It lies prostrate, as perfect a picture of weed death as one would care to see. Then, the next morning, all the finger-tipped branches are standing upright and investigation shows that not only the single root has done its work but that nearly every joint on its prostrate stems is now sending roots into the ground. It is a weed of nine lives. It is the sanguine finger grass.

But we have to admit that the weed has a few virtues. The young plants serve well as a fertilizer, and much of the garden-patch fertility is due to the Crab grass seedlings that are daily being hoed into its soil.

Many strawberry growers have found that Crab grass allowed to grow and die in their strawberry rows serves as a substitute for the straw that must needs be spread on grass-free rows. The grass protects the plants through the winter, and, when the berries set in the spring, the layer of dead grass keeps the berries out of the dirt.

Crab grass is not a substitute for bluegrass, nor can any annual grass take the place of the perennial when it comes to lawn making, but many a poorly sodded lawn has been made to appear perfect when Crab grass came in and filled the vacant spaces. It is a lawn pest if the lawnmower is not kept on the job, but constant clipping makes of it a substitute for the lacking bluegrass throughout the growing season. Of course the patches are left bare when winter comes.

In these three ways then, as a fertilizer, as a mulch, and as a substitute for lawn grass, the weed makes some restitution for its weediness.

TECHNICAL DESCRIPTION

Digitaria Scop. Spikelets 1–flowered, lanceolate-elliptic, sessile or short-pediceled, solitary or in 2's or 3's, in two rows on one side of a continuous narrow or winged rhachis, forming simple slender racemes which are aggregated toward the summit of the culm; glumes 1–3-nerved, the first sometimes obsolete; sterile lemma 5–nerved; fertile lemma leathery-indurated, papillose-striate, with a hyaline margin not inrolled, inclosing a palea of like texture.—Annual, mostly weedy grasses, with branching culms, thin leaves, and subdigitate inflorescence. (Name from *digitus*, a finger.)

Digitaria sanguinalis (L.) Scop. Culms erect or ascending from a decumbent often creeping base, 3–12 dm. long; nodes and sheaths more or less papillose-hirsute; blades lax, 5–12 cm. long, 4–10 mm. wide, scabrous, often more or less pilose; racemes 3–12, subfasciculate, 5–18 cm. long; *spikelets* in pairs, 3–3.5 mm. long, usually appressed-pubescent between the smooth or scabrous nerves; second glume about ½ as long as the *pale* or grayish fertile lemma.—Cultivated and waste grounds, throughout our range, and southward. August–October. Very variable. (Naturalized from Europe.)

FALL PANIC GRASS

[*Panicum dichotomiflorum* Michx.]

TICKLE GRASS, Old witch grass, Spreading witch grass, Panic grass and, as the botanist says, *Panicum dichotomiflorum,* are likely to be all the same thing. There are a great many panic grasses in this country, however. Hitchcock in his *Manual of the Grasses of the United States* describes 160 species. Many of these are recognized only by the botanist who is specializing in grasses, but this one may be recognized by any one who takes the trouble to look at it and see that in the great bunch of reclining grass stems which come from the central root system each stem bears two or more panicled branches filled with flowers or seed. If the observer has had to plow under such clumps of grass he is almost sure to have had to stop his plow occasionally to pull out a particularly bad clump which caught between the plow and the colter. Ten to one that bunch of grass was the Fall panic grass.

Of course the name panic refers to the Latin name, *Panicum. Panicum* was the old Latin name for millet. *Dichotomiflorum* refers to the mode of flowering. Dichotomy means forking with equal branches. *Dichotomiflorum* means that the flowers are in twos on equal branches. They are not always on equal branches nor are they always in twos, but the man who named the grass (André Michaux) either thought he saw or did see that most of the flowers were so arranged that two little branches of equal length would each bear a flower, and so he called it *Panicum dichotomiflorum.*

The grass is a real weed in some localities. It will take a meadow if it gets a chance, and in rich sour ground it can crowd out almost any of the cultivated crops. It is an acid tolerant plant, and so has the advantage over any plant that is sensitive to soil acidity. Farmers often call this weed, along with several other grass species, "sour grass" and "swamp grass." It is rather fond of wet places but it is not a swamp grass, nor is it a sour

FIG. 4. The Fall Panic Grass. *Redrawn
from Muenscher's "Weeds"*

grass if the term means that it will sour the soil when it is used as a fertilizer. It actually makes a very good fertilizer if plowed under not later than the blooming stage. Like all other succulent weeds it can be and should be used in this way as a soil builder.

TECHNICAL DESCRIPTION

Panicum L. Spikelets 1-flowered or rarely with a staminate flower below the terminal perfect one, in panicles, rarely in racemes; glumes very unequal, the first often minute, the second subequal to the sterile lemma which often incloses a hyaline palea and rarely a staminate flower; fertile lemma and palea chartaceous-indurated, nerves obsolete, the margins of the lemma inrolled; grain free within the rigid firmly closed lemma and palea.—Annuals or perennials of various habit. (An ancient Latin name of the Italian millet, *Setaria italica,* of uncertain origin and meaning.)

Panicum dichotomiflorum Michx. Culms compressed, thick, succulent, spreading or ascending from a decumbent base, 3–18 dm. long; leaves 2–4 dm. long, 8–15 mm. wide, scabrous above; panicles 1.2–4 dm. long, diffuse; spikelets short-pediceled, mostly secund toward the ends of the branchlets, 3 mm. long, acute; first glume obtuse, second and sterile lemma pointed beyond the fruit.—Low waste grounds and cultivated fields, Maine to Nebraska, and southward. July–October.— Slender, depauperate, erect or prostrate specimens occur in sterile ground.

BARNYARD GRASS

[*Echinochloa crusgalli* Beauv.]

IF YOU want to impress your hearers with your erudition learn to say *Echinochloa crusgalli* (ĕk-ĭ-nŏk'lō-à crŭs-găl'-lĭ). Those mouth-filling words mean the Cockspur, Hedgehog grass. The first word, *Echinochloa,* is made up of two words: *Echino* is from the Latin and means hedgehog, and *chloa* from the Greek means grass. The crusgalli part is made up of two Latin words: *crus* meaning shank, and therefore the place, at least, where the spur grows, and *galli,* which is a form of the word *gallus,* meaning cock. These two words show how the weed has impressed its namers. Linnæus saw the cockspur resemblance in the awns, and A. M. F. J. Palisot de Beauvois (that is actually his name) saw the resemblance to the hedgehog in the entire head of the grass.

The weed has several common names also. Besides being called Barnyard grass, because it is usually found in barnyards, it is called Water grass, Panic grass, Cockspur grass, Cockfoot grass, Barn grass, and Panicum. It grows all over the United States and Canada, wherever the soil is rich and wet enough, except in the extreme North It is a weed of the first water, even though it is occasionally used for hay along with some of the other wild grasses. At its best it makes a poor hay, and at its worst it will take a corn or cotton field, a feed lot or a pasture. It is an annual that seeds heavily and depends upon its lithe stems of from two to six feet in height to scatter its seed.

This grass is so different from any other that one can quickly learn to know it at sight. It is a late blooming weed and so does not attract attention much before the latter part of the summer. Because of this fact it seldom bothers the lawn owner, but if he has a pool or a low place in some part of his yard and is neglectful enough to permit the grasses to grow up there, it is nine chances to one that he will make the acquaintance of Barnyard grass before the season closes. It is a lover of rich wet soil. The richer the soil the better. That is why the weed frequents barnyards. There it finds the rich manure deposits that are too rich for other weeds. If you do not know the weed but you do know where there is a manure pile in a barnyard, take a look at the grasses near that pile in late summer and see if you do not find *Echinochloa crusgalli* holding forth there.

TECHNICAL DESCRIPTION

Echinochloa Beauv. Spikelets 1-flowered, sometimes a staminate flower below the perfect terminal one, nearly sessile in 1-sided racemes; glumes unequal, spiny-hispid, mucronate; sterile lemma similar and awned from the apex (sometimes mucronate only), inclosing a hyaline palea; fertile lemma and palea chartaceous, acuminate; margins of

FIG. 5. The Barnyard Grass is so different
it is easily identified

the glume inrolled except at the summit, where the palea is not included.—Coarse annuals with compressed sheaths, long leaves and terminal panicles of stout racemes. . . .

Echinochloa crusgalli (L.) Beauv. Culms stout, rather succulent, branching from the base, ascending or erect, 3–18 dm. high; *sheaths and blades glabrous;* panicle dense, 1–3 dm. long, of numerous erect or spreading racemes, very variable, deep purple to pale green, erect or drooping; spikelets long-awned or nearly awnless, densely and irregularly crowded in 3 or 4 rows, about 3 mm. long. Moist, chiefly manured soil and waste ground, river banks, etc., common throughout, except in the extreme North. August–October. (Naturalized from Europe.)

YELLOW FOXTAIL

[*Setaria lutescens* Hub.]

Few people who know cornfields in the Mississippi Valley have failed to form an intimate acquaintance with this weed. It is not a bad weed if the farmer is a good farmer and cultivates his ground sufficiently. But the lazy farmer is the friend of the Foxtails. They fill every available inch of neglected cornfield if the Panicums do not beat them to it.

The name foxtail is most fitting. The head, or spike, looks very much like a miniature fox's tail. It has the color and it bristles out very much as a fox's tail does. The botanical name, *Setaria lutescens,* means yellow bristle grass. The *Setaria* comes from the Latin word *seta,* which means a bristle, and the *lutescens* (pronounced lū tĕs′ cens) means yellowish. In different localities it is called Wild millet, Golden foxtail, Foxtail, Summer grass, and Pigeon grass, but yellow foxtail is what it should be called to distinguish it from its cousin, the green foxtail.

The weed is a pest for two reasons: first, it grows very rapidly, and second, it does not make good food. It is not eaten after it fruits by either cattle or horses. The only use it seems ever to have been put to is that of a band for tying up corn fodder. A bunch of the culms (stalks) of this grass twisted together makes a very good rope.

Fig. 6. The Yellow Foxtail is well named

TECHNICAL DESCRIPTION

Setaria Beauv. Spikelets as in *Panicum* but surrounded by few or many persistent awn-like branches which spring from the rhachis below the articulation of the spikelets.—Annual introduced weeds in cultivated or manured grounds, or native perennials, with linear or lanceolate flat leaves and cylindrical spike-like panicles. (Name from *seta,* a bristle.)

Setaria lutescens Hubb. Annual; culms branching at the base, compressed, erect or ascending, 3–12 dm. high; leaves flat, linear-lanceolate, glaucous; panicle 2–10 cm. long, about 1 cm. thick; *bristles* 3–8 mm. long, *upwardly scabrous; spikelets* 3 mm. *long; first glume* ½, *second* ⅔ *as long as the striate undulate-rugose fertile lemma.*—Cultivated ground and waste places, common throughout. (Naturalized from Europe.)

SANDBUR

[*Cenchrus tribuloides* L.]

THIS pest of sandy soil is seldom found anywhere else, but wherever there is sand, and along highways and parkways near sandy areas, it is almost sure to put in its hateful appearance. Most people know only the bur, which is one of the sharpest of armed fruits. It sticks to the clothing and pierces the fingers when one attempts to remove it. No other grassy plant has this fruit, this lance-armed bur, and a description therefore is almost useless. He who finds a sand bur does not have to be told what it is.

Before its burs are ripe the plant is a grass growing from three to eight inches in length and its stems lie spread out on the ground. There are many branches to each plant with racemes of burry flowers at the end of each branch. It fruits from June through September.

The botanical name of the plant is *Cenchrus tribuloides.* *Cenchrus* is the ancient Greek name for a particular kind of millet, and the *tribuloides* probably refers to the emotional reaction of him who steps on the bur with his bare foot.

FIG. 7. The hateful Sandbur

TECHNICAL DESCRIPTION

Cenchrus L. Spikelets 1–flowered, acuminate, 2–6 together, subtended by a short-pediceled ovoid or globular involucre of rigid connate spines which is deciduous with them at maturity; glumes shorter than the lemmas; sterile lemma with a hyaline palea, fertile lemma and palea less indurated than in *Panicum,* falcate-acuminate, the lemma not inrolled at the margins.—Our species annual, with simple racemes of spiny burs terminating the culm and branches.

Cenchrus tribuloides L. Culms more robust, *often extensively branching or trailing,* 3–9 dm. long; sheaths loose, usually hirsute along the margins, ligule conspicuously ciliate; *blades more or less involute;* racemes usually included at the base; *involucres 12–14 mm. thick, densely long-pubescent; the stout spines spreading or ascending.*— Sands along the coast, New Jersey and southward.

NEEDLE GRASS

[*Aristida oligantha* Michx.]

To GET well acquainted with needle grass one has only to walk through a patch of it in the late fall of the year. Unless his legs are in puttees he will forever after know what needle grass is. It is not much of a weed except where the ground is poor. In sour ground it may become as thick as wool and as menacing as a thicket of thistles. Ordinarily it amounts to very little. It is seen in small bunches, its thin stems and blades setting it off from the rest of the vegetation as something of a weakling in the grass line. It is an annual and so never makes a permanent sod, and although it is actually "gone with the wind" soon after the seed is ripe, the wind that takes it leaves it in every available niche. For this reason needle grass is likely to appear in any lawn, pasture, or meadow, but if the soil is a good soil the weed will never be noticed. If the soil is sour or in any way intolerant to the sod-making grasses the door is left open for the needle grass and it is sure to enter in all its worthless, prickling glory.

The grass has the following common names: Prairie three awn

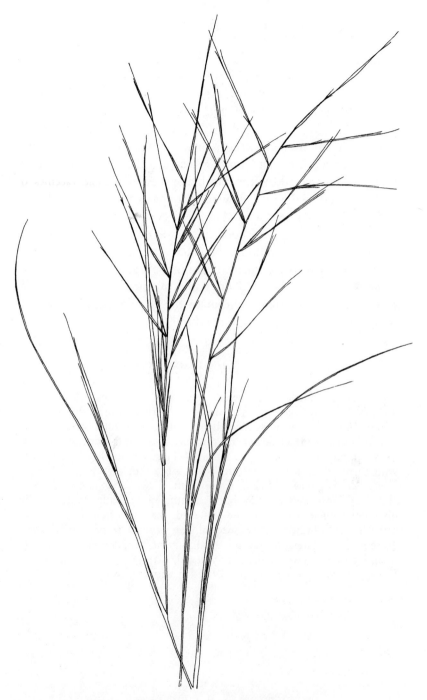

FIG. 8. Needle Grass or Three-awned grass

grass, triple awn grass, few flowered aristida and wire grass. The last name is the only one except needle grass that has come from the people. All the other names are pseudoscientific. The botanical name, *Aristida oligantha* Michx., means the little flowered awn grass. *Aristida* (pronounced aris'tida) is Greek for awned or armed with awns. *Oligantha* is made up of two Greek words, *oligo* meaning little and *anthus* meaning flower. So *oligantha* means little flowered. A free translation then of *Aristida oligantha* Michx. would be the little flowered awn grass named by Michaux.

From what has been said above it is evident that the best way to eradicate needle grass is to keep the soil of lawns, pastures, and meadows in good condition for good grasses. A little lime and a little nitrogen in the form of sodium nitrate, ammonium sulphate, cyanamid, or just plain stable manure spread on in the winter time will usually eliminate the needle grass. It cannot compete with other grasses in good soil.

TECHNICAL DESCRIPTION

Aristida L. Spikelets 1-flowered, in usually narrow panicles; glumes unequal, narrow, acute or acuminate; a hard obconical hairy callus below the floret; lemma somewhat indurated, convolute, including the thin palea and perfect flower, terminating in a trifid awn; grain elongated, tightly included in the lemma.—Tufted annuals or perennials with narrow leaves. (Name from *arista,* a beard or awn.)

Aristida oligantha Michx. *Culms* tufted, wiry, *branched at base and at all the nodes,* 3–6 dm. high; sheaths loose; blades long, usually involute; panicle or raceme few-flowered, the axis often flexuous and spikelets spreading; *glumes* unequal, *long-awned from a bifid apex,* exceeding the floret, the second strongly 7-nerved; *lemma* 17–20 mm. long, scabrous above; awns nearly equal, divergent, 3.5–7 cm. long.— Dry sterile soil, New Jersey to Nebraska, and southward.

NIMBLE WILL

[*Muhlenbergia Schreberi* J. F. Gmel.]

NIMBLE WILL is one of the most descriptive names a plant has ever received *viva voce*. If any one with a slight interest in plants was told that one of the grasses is called Nimble Will, he would know that this was the grass as soon as he had made its acquaintance. And yet few people know Nimble Will. In fact, few people have names for any of the grasses. Even Crab grass with all of its aggressiveness is just another d—— weed in the minds and mouths of most of the profane men with hoes.

Nimble Will is nothing like as aggressive as Crab grass, but it is a weed and one that every lawn owner should know. It is the kind of weed that promises much in the early stages and then turns out to be as worthless and as full of seed as a hillbilly of the movies. The roots are perennial but the stems are not in the least so. They are dry and ready to burn as soon as the first frost of autumn arrives.

The long, slender stems (culms) which have a tendency to take root at the lower nodes help to identify the plant. These slender, wiry stems, often from two to three feet in length, recline more or less and so make "Limber Will" quite as applicably descriptive as "Nimble Will." The "nimble" of the name probably referred originally to the grass's ability to catch, or was it because we think of limber people being nimble people? Anyway, Nimble Will is a limber grass.

The habitat of Nimble Will, according to Gray's *Manual of Botany,* is "Dry woods, hillsides and waste places." The "waste places" are likely to be an orchard or a poorly kept lawn. Even a well-kept lawn may have Nimble Will in it. The young grass will satisfy the keeper, but when fall comes there will be brown, bare spots where deception left its marks, and under the brown spots will be the perennial roots of Nimble Will.

The weed is also called Drop-seed and Wire grass. The latter

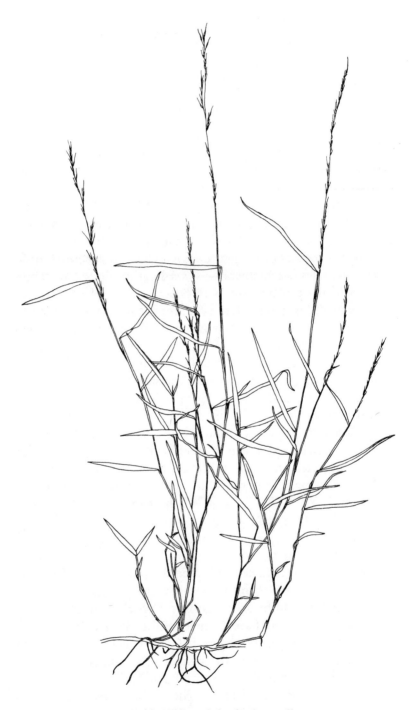

Fig. 9. Nimble Will with his limber stalks

name is descriptive enough, but it so happens that several grasses deserve the name of wire grass while few, if any, of them could be called Nimble Will. The botanical name is likely to be a poser for him who does not realize that the most of the grasses have been named quite recently, and that distinguished botanists have been attempting to honor their distinguished friends by giving to some of the unnamed plants the names of some of these friends. Such is the case of Nimble Will. Both the generic and the specific names are the modified names of men. The genus, *Muhlenbergia,* was named in honor of Doctor Henry Muhlenberg, a distinguished American botanist. The specific name, *Schreberi,* is given in honor of the botanist who suggested *Muhlenbergia,* Johann D. C. von Schreber. So *Muhlenbergia Schreberi,* J. F. Gmel. distinguishes the plant with distinguished names, for the distinguished John Friedrich Gmelin suggested the name *Schreberi.* What a load of distinction poor Nimble Will has to carry!

TECHNICAL DESCRIPTION

Muhlenbergia Schreb. Spikelets 1–flowered, in contracted (rarely open) panicles; a short usually barbate callus below the floret; glumes thin, often aristate; lemma narrow, membranaceous, 3–nerved, awned or awnless, inclosing a thin subequal palea; grain closely enveloped by the lemma.—Our species perennial, often with scaly rootstocks, flat or involute leaves and small spikelets. (Dedicated to the *Reverend Doctor Muhlenberg,* a distinguished American botanist, 1753–1815.)

Muhlenbergia Schreberi J. F. Gmel. Culms 3–8 dm. long, erect or ascending from a decumbent base, often rooting at the lower nodes, diffusely much branched; blades 3–8 cm. long, 2–4 mm. wide; panicles 5–15 cm. long, numerous, slender, the erect branches rather densely flowered; spikelets (excluding the awn) 2 mm. long; *first glume obsolete or nearly so, the second minute, truncate;* lemma tapering into a slender awn 3–5 mm. long. (*M. diffusa* Schreb.)—Dry woods, hillsides and waste places, Maine to Ontario, Minnesota, and southward. August–September.

BERMUDA GRASS

[*Cynodon dactylon* Pers.]

No ONE in the South has to be told what Bermuda grass is, but some Southerners should be told that it is not entirely a weed. That it is bad under certain conditions the most ardent advocate of the plant has to admit. It can take a cornfield or a cotton patch, and the farmer who hopes completely to eradicate it from his cultivated fields can expect to fight throughout a growing season, and be ever on the lookout every season thereafter.

But Bermuda grass has virtues as well as faults. It is the peer of grazing grasses. It has no equal as a pasture grass, especially in the South. Its range is much farther north than grass-growing farmers are likely to believe. Pastures and lawns can be made of it as far north as St. Louis, Missouri. It has two faults as a lawn grass: it starts late in the spring and it refuses to grow in the slightest shade. Of course, the lawn owner wants a grass that will shoot up green as soon as the growing season starts. The Bermuda grass will not do this, but in July and August when other lawn grasses have to be pampered the Bermuda grass is luxuriant.

It is in the pasture that the Bermuda grass excels, however. A good field of Bermuda will pasture five head of cattle to the acre throughout the growing season. There are two varieties, or perhaps species, of Bermuda grass found in the South. The tall or large variety can be used as a hay grass. A fifteen-acre field of this grass at Fort Smith, Arkansas, has been growing a ton of grass to the acre for several years and the hay made from it is said to be very nutritious.

For the information of him who does not know Bermuda grass it should be said that its general appearance is somewhat like that of Crab grass. It has the fingerlike fruiting top that the Crab grass has, but the fingers are smaller and much shorter than those of the Crab grass. They also extend out horizontally, almost perpendicular to the supporting stem. The leaves are not

FIG. 10. Bermuda Grass, one of the
best of pasture grasses

nearly so long as those of the Crab grass, and much finer in texture, and there are runners sent out from the main bunches such as are never seen in any other grass except perhaps the Buffalo grass. But the runners of Bermuda grass are much heavier and longer than those of the Buffalo grass. If you find in your yard a thin-stemmed, short-leafed grass that is difficult to cut with a lawn mower, and if you find that it has a fruiting top similar to that of Crab grass you may know you are in possession of the grass from Bermuda—if it came from there.

The prejudice against Bermuda grass has all come about because the southern farmer does not know how to keep it out of his cultivated fields. This is easily done by shallow plowing of the field just before the ground freezes in late autumn or early winter. After the ground freezes the Bermuda is easily killed. This one plowing, which turns the rootstalks up on top of the ground, and so destroys them by freezing, leaves little to be done except the essential cultivation of the next year. Near Van Buren, Arkansas, is a truck farm producing the finest kind of cabbage, tomatoes, and potatoes and this farm was made out of a field solidly covered with Bermuda grass. The Bermuda was destroyed in a single year, by the method given above.

The feeling against the Bermuda is so great that farmers in the South refuse to allow a field to produce Bermuda hay. They fight Bermuda all summer and buy hay all winter. Only the dairymen of the South seem to appreciate fully the value of this wonderful weed. Ten or twelve acres of Bermuda grass are the equivalent of five or six times that acreage of bluegrass or red top and timothy.

Because of its value as a hay, pasture, and lawn grass, at least a paragraph should be devoted to its use and propagation. It is possible to buy and sow Bermuda grass seed, but few dirt farmers believe it is possible to get a stand of grass in that way. They declare that only in the most southern parts of the southern States will seedling Bermuda grass stand the first winter. The very best way to insure a set of the grass is to plant it. The runners are plowed up and chopped into bits and scattered into shallow furrows in the field made ready for setting. The runners should be kept wet, and as fast as the pieces of them are dropped in the furrows they should be covered two or three inches deep, and as soon as possible after that the ground should be packed with a roller. This is the way to get a good lawn or a good pasture or hayfield.

The grass grows very slowly the first year but after it has established itself, especially south of the Mason and Dixon line, it can be expected to produce good growth in the second season. It responds readily to lime-filled soils and oftentimes an application of limestone will act like a top dressing of manure to a dwindling Bermuda pasture. It is one of the very best plants to use for controlling soil erosion. Ditches may be actually filled by planting a few bunches of Bermuda grass in the bottoms of them. It will hold dams for impounding water almost as effectively as a concrete core and it can be used very effectively on terrace outlets or spillways. There are so many places where Bermuda grass will grow and serve the landowner that it is almost sacrilege to call it a weed, but what else is it when the enraged farmer finds that it has choked large areas of his cotton plants to death?

The weed has several names besides the one most commonly used. It is called Wire grass—a good name for it—Scutch grass, Dog's Tooth grass, Bahama grass, and Devil's grass. Its botanical name, *Cynodon,* is from two Greek words meaning dog and tooth. Its specific name, *dactylon,* means fingerlike. It is, therefore, the dogtooth grass with fingerlike fruiting heads.

TECHNICAL DESCRIPTION

Cynodon Richard. Spikelets 1-flowered, laterally compressed, awnless, singly sessile in 2 rows along one side of a slender continuous axis, forming unilateral spikes; rhachilla prolonged behind the palea into a blunt pedicel; glumes unequal, narrow, acute, keeled; lemma broad, boat-shaped, obtuse, ciliate on the keel; palea as long as the lemma, the prominent keels close together, ciliolate; grain free within the lemma and palea. Low diffusely branched and extensively creeping perennials, with flat leaves and slender spikes digitate at the apex of the upright branches.

Cynodon dactylon (L.) Pers. Glabrous; culms flattened, wiry; ligule a conspicuous ring of white hairs; spikes 4-5, 2-5 cm. long; spikelets imbricated, 2 mm. long; lemma longer than the glumes.—Fields and waste places. Massachusetts and southward, where it is cultivated for pasturage. (Naturalized from Europe.)—Seldom perfects seed.

GOOSE GRASS

[*Eleusine indica* Gaertn.]

GOOSE GRASS is often confused with two other grasses: Crab grass and Crowfoot grass. It is even called Crowfoot grass, but there is a difference between the true Crowfoot species and Goose grass that even the layman can quickly detect. The fingerlike seed heads of Goose grass are blunt at the ends; the Crowfoot grass fingers end in naked stems, much like claws. The visible difference between Goose grass and Crab grass is more in size than in anything else. The seed-bearing fingers of the Goose grass are heavy and thick in comparison with the slender, long fingers of the Crab grass.

Moreover, the habitat of the Goose grass is different from that of nearly all of the grasses. It is more like the habitat of the knotweed and the wire grass (Juncus tenuis). The weed likes to grow where the ground is packed hard. It will grow in a not too frequently used path. It likes cinder beds, if they are packed, and it can become a serious lawn weed where the ground is poor and hard. The weed grows all over the eastern part of the United States from Nebraska to the Atlantic coast, and it is also found in California and Oregon.

The seed, or grain, of the weed is so large that it was at one time used to make a poor grade of flour. In some of the poorer sections of Europe and Asia it is said that Goose-grass flour was made (perhaps still is made) from a selected variety of the weed. The fact that it thrives in poor, hard soils makes of it a desirable grain plant where such soil is the only soil. The generic name of the grass, *Eleusine,* suggests a place of very poor soil. It is the name of the Greek town where Ceres, the goddess of grains, was worshipped.

Fig. 11. The Goose Grass has a seed from
which flour may be made

The species name, *indica*, probably refers to the place of origin of the grass: India. Linnæus, however, placed the plant in the genus *Panicum* and called it *indicum*, and since *indicum* is Latin for indigo he might have given it this name because of the bluish cast the plants have. The entire name, as it now stands, *Eleusine indica*, was given by Joseph Gaertner.

It should be mentioned that some people call Goose grass Wire grass, and others call it Yard grass. Any tough grass is likely to be called wire grass, and any grass that frequents a back yard is just as likely to be called yard grass; but so few grasses deserve these names that one can be sure, within a range of two or three grasses, what is meant when either name is used.

TECHNICAL DESCRIPTION

Eleusine Gaertn. Spikelets several-flowered, awnless, florets perfect or uppermost staminate, sessile and closely imbricated in 2 rows along one side of a continuous rhachis, which does not extend beyond the terminal spikelet; glumes unequal, shorter than the floret, scabrous on the keels; lemmas broader, with a thickened 5-ribbed keel; palea shorter, acute, the narrowly winged keels distant; grain black, the loose pericarp marked with comb-like lines, free within the subrigid lemma and palea.—Coarse tufted annuals with stout unilateral spikes digitate or approximate at the apex of the culms. (Name from the Greek town where Ceres, the goddess of harvests, was worshipped.)

Eleusine indica Gaertn. Glabrous; culms flattened, decumbent at base; sheaths loose, overlapping, compressed; spikes 2-10, 2.5-8 cm. long; spikelets appressed, 3-5-flowered, about 5 mm. long.—Yards and waste ground, Massachusetts, northern Illinois, Kansas, and southward. (Naturalized from tropics of the Old World.)

TALL RED TOP

[*Tridens flavus* Hitchc.]

THERE is a tall red-topped grass that every one who sees anything in a late summer or early fall landscape is sure to see. It has beauty enough in grace and color to be gathered for bouquets, but it is

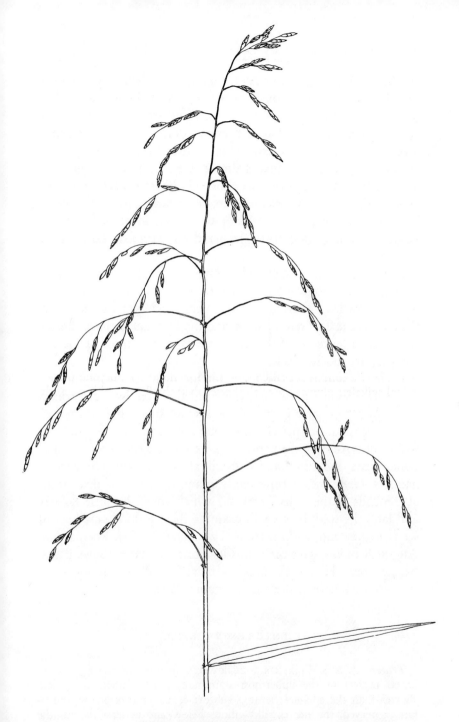

Fig. 12. The Tall Red Top is
beautiful but worthless

only a weed. Stock almost never eat it; never after its beautiful panicle of flowers is formed, for there is a viscid substance that issues from the branches of the panicle and from the stem below it, and that substance has a strange odor. It is the odor that gives the tang to the evening air of that time of the year; a peculiar, Oriental smell that is almost entrancing to some nostrils, but is evidently disgusting to those of grazing stock. For this reason tall red top is on the increase. Nearly every roadside is lined with it now, and it is likely to be seen in partially grazed pastures, in neglected town lots and in unmown farm fence rows.

The weed is a perennial and its bunches of long, narrow leaves distinguish it from most other grasses before it sends up its panicle-crowned stems (culms). Very few grasses grow in bunches as the Tall red top does, and the few that do have shorter and broader leaves, or if the leaves are as long as those of the red top, they are much broader.

Tridens flavus is a good name for the plant. It means the yellow, three-toothed grass. It is more red than yellow, but perhaps Hitchcock, who gave it that name, saw a variety that was not so red—there is one—or perhaps this color appeared more yellow than red to him. Anyway the name is *flavus* and not *rubrus* or *rubrum,* whichever it should be, and so it will always remain. The generic name, *Tridens,* means three-toothed, and refers to the three teeth at the ends of the hulls (lemmas) which cover the little flowers and later the seed. If one will examine these hulls closely he will see that they end in three teeth. Two botanists, J. J. Roemar and August Schultes, gave the genus its name. As stated above, it was Albert Spear Hitchcock, another botanist and an authority on grasses, who gave it the specific name, *flavus.*

TECHNICAL DESCRIPTION

Tridens R. & S. Spikelets 3–12-flowered in open or strict panicles; florets perfect or the uppermost staminate; glumes unequal, keeled, shorter than the spikelet; lemma subcoriaceous, convex below, bidentate, 3–nerved, the nerves silky-villous below and at least the middle one extending in a mucronate point between the teeth; palea broad, the nerves nearly marginal.—Perennials with long narrow leaves and terminal panicles. (Name from *tres,* three, and *dens,* tooth.) *Tridens flavus* (L.) Hitchc. Culms erect, 1–2 m. high, viscid in the axis of the panicle and below it; sheaths bearded at the summit, otherwise glabrous as are the long flat or involute tapering blades; the *showy panicles 2–4.5 dm.* long, *almost as wide, loose and open, the slender branches spreading, naked below; spikelets purple,* 7–8 mm. long, 5–8-flowered, on long pedicels; *glumes shorter than the lowest florets, mucronate; the three nerves of the lemmas excurrent.* Dry or sandy fields, Connecticut to Missouri, and southward. August–September.

CHEAT

[*Bromus secalinus* L.]

DOES WHEAT turn to cheat? In spite of agricultural colleges and farm advisers that question is still being asked and still being answered both ways—not by the educated farmer, of course not. He knows that the answer is no, but that "no" makes no impression on the man of experience who has seen his field of wheat freeze out and produce a fine field of cheat.

At any rate Cheat, Chess, Wheat thief, Smooth rye grass or Cock grass is one of the worst weeds of the wheat field. Wheat does not turn to cheat. So long as the wheat is rank the cheat's slender and much shorter stalks are not noticed. They are usually there, however, and as sure as the wheat fails the cheat will take its place. How did the cheat get there? Well, one seldom finds wheat seed that is free from cheat seed. The cheat is often sown right with the wheat. The weed's seed is ripe when the wheat is, and no matter how badly smothered out by the rank growth of wheat there will be places where the cheat heads will develop and enter the threshing machine along with the grain. Then, to remove that cheat seed from the wheat seed is so much of a task that it is seldom accomplished. And even if it is done, and well done, the weed has another string to its bow. It does not try to

Fig. 13. Cheat or Chess

escape observation, but it actually does escape it. It is an insignificant, harmless-looking grass in a fence row or in a pasture or even in a truck patch. Just a bunch of grass that is harmless—harmless until those seeds it is developing are scattered in a near-by wheat field, and some of them are sure to land there, when Mother Nature's seed-sowing winds pass that way.

In other words, it is nearly impossible to keep cheat out of wheat. It is found in winter wheat wherever it is grown in the United States and Canada. It does not bother spring wheat fields, but there is probably not a State in the Union where the weed does not grow.

Bromus, the generic name, is the ancient Greek name for oats; it really means food. *Secalinus* is derived from the generic name of rye, *Secale.* Secalinus refers to the general appearance of the grass. *Bromus secalinus* would mean the Bromus that looks like rye.

TECHNICAL DESCRIPTION

Bromus L. Spikelets few–many-flowered; glumes unequal, acute, 1–5-nerved; lemmas longer than the glumes, convex or sometimes keeled, 5–9-nerved, usually 2-toothed at the apex, awnless or awned from between the teeth or just below; palea a little shorter than the lemma, 2-keeled; grain furrowed, adnate to the palea.—Annuals, biennials, or perennials with flat leaves and terminal panicles of rather large spikelets. (An ancient name for the oat, from the Greek word meaning food.)

Bromus secalinus L. Culms 4–9 dm. high; *sheaths smooth and strongly nerved;* blades sparingly pilose above; *panicle open, its branches somewhat drooping;* spikelets 5–15 flowered, glabrous; glumes 5–7 mm. long; lemma 8–11 mm. long, becoming at maturity convex, thick and inrolled at the margins, awns short and rather weak.—Fields and waste places, common.—The florets are somewhat distant, so that, in side view, openings are visible along the rhachilla at the base of the florets. (Naturalized from Europe.)

SQUIRREL-TAIL GRASS

[*Hordeum jubatum* L.]

THE BEAUTY of the shining heads and swaying grace of Squirrel-tail grass in a border along a highway or path is convincing proof that beauty and worth are occasionally divorced. Beautiful it may be, but worthless wholly. Grazing animals will eat a little of it when it is young, but they give it a wide berth after its head is formed, and hay containing Squirrel-tail grass is scarcely fit to feed. The beards (awns) penetrate the gums, lips, tongues, and even the eyelids of the stock feeding on such hay.

The weed is such a frail little thing that one is likely to disregard it when a single bunch of it appears in a pasture or on a parkway near one's home, but as sure as it appears it will stay, and as sure as it stays its tribe will increase, unless something is done about it. If nothing is done about it those who walk that way will soon be cursing the pest and spreading its seeds with their socks and stockings. That is what those beautiful tails are for. Every bristle on them is barbed and carries at its lower end a seed. The beards pierce the passing stockings, the barbs hold them there, and the seeds ride to new locations where the owner of the stockings dislodges the stinging awns.

Here is one weed that did not have to come through Old England to acquire its several names. It is a native (according to some authorities) and yet besides being called Squirrel-tail grass it is called Skunk-tail grass (which suits it much better than Squirrel-tail), Flicker-tail grass, Wild barley, which it is, Tickle grass, which it is not, and Foxtail grass, which again it is not. Tickle grass should never be applied to anything except *Panicum capillare*. That is the grass that actually tickles. It is the one whose finely divided top starts in at the bottom of a trouser leg and crawls out at the collar band, if the victim is not sufficiently tickled to stop the ticklers on their joyful climb. And foxtail belongs to the *Setarias, Setaria lutescens* or *Setaria viridis*. They are the true foxtails.

Fig. 14. **Squirrel-tail Grass,** one
of the wild barleys

The generic word, *Hordeum,* is the ancient name for Barley and *jubatum* is from the Latin, *jubatus,* meaning maned. So Linnæus scored again when he called this plant the barley with a mane.

TECHNICAL DESCRIPTION

Hordeum (Tourn.) L. Spikelets 1 (rarely 2)–flowered, 3 together in our species at each joint of the flattened articulate rhachis, the middle one sessile, perfect, the lateral pair usually pediceled, often reduced to awns and together with the glumes of the perfect spikelet simulating a bristly involucre at each joint of the rhachis; rhachilla prolonged behind the palea as an awn, sometimes with a rudimentary floret; glumes equal, rigid, narrow-lanceolate, subulate or setaceous, placed at the sides of the dorsally compressed floret which is turned with the back of the palea against the rhachis of the spike; lemma obscurely 5–nerved, tapering into an awn; palea slightly shorter, the 2 strong nerves near the margin; grain hairy at the summit, usually adherent to the palea at maturity.—Cæspitose annuals or perennials with terminal spikes which disarticulate at maturity, the joints falling with the spikelets attached. (The ancient Latin name.)

Hordeum jubatum L. Biennials, 3–7 dm. high, erect or geniculate at base; leaves 5 mm. wide or less, scabrous; *spike nodding, 5–12 cm.* long, about as wide; lateral pair of spikelets each reduced to 1–3 spreading awns; *glumes* of perfect spikelets awn-like, 3–6 cm. long, spreading; lemma 6–8 mm. long, with an awn as long as the glumes; all the awns very slender, scabrous.—Coast, Labrador to New Jersey; prairies and waste ground, Ontario to Illinois, Kansas, and westward. June–August.—Often a troublesome weed. (Eurasia.)

WILD BARLEY

[*Hordeum nodosum* L.]

THERE are several wild barleys, but this one is perhaps the worst weed of the lot since it is the one most capable of taking over pastures and meadows. It is sometimes called Meadow barley because of its meadow-snatching proclivities. It is a relative of Squirrel-tail grass, *Hordeum jubatum,* and has the same habitats *jubatum* has. It is less conspicuous than the Squirrel-tail, but more aggressive, and never fails to seize upon every spot left open to it.

In most places the grass is an annual, though it is said to be

FIG. 15. Wild Barley, the pest of pastures

both annual and perennial. It gives a great deal of promise when it appears, and it appears early in the spring. Grazing stock feed upon it in its early stages, and one of the best ways to fight the weed is to graze it off and thus prevent its seeding. The plant is too small to crowd out taller weeds and crop plants, but it is acid-tolerant and so can grow in soil too sour for most plants. It cannot crowd out timothy, bluegrass or alfalfa, but it quickly crowds in where such crops are struggling against sour soil. When a meadow or a pasture becomes infested with *Hordeum nodosum* it should be plowed, limed, and reseeded after a cultivated crop or two have been grown there to insure the removal of the last vestige of the barley.

It promises much. Many farmers and lawn-owners have watched what they thought was a fine set of bluegrass or timothy and found it to be only this evanescent pest. It soon heads and seeds and then the heads (spikes) break up into as many parts as there are spikelets on them, and since the spikelets are armed with awns that attach themselves to every pair of trouser legs and stockings that pass by, the seed is often widely scattered in this way. But the wind is the agent of distribution most depended upon. A gust of wind or a whirlwind will break up the ripe heads and send the spikelet-bearing seeds over the fences and fields to new conquests. That is why it is "widespread in the middle western states and to the Pacific Coast."

This particular species, *nodosum,* might be, and perhaps often is, confused with another species called *Hordeum pusillum.* The taxonomists do not seem to be wholly in agreement as to the differences between the two, and so laymen could hardly be expected to tell the two apart. *Pusillum* is supposed to be smaller than *nodosum,* to be always an annual and to live in alkaline soils, while *nodosum* is tall, at least from ten to twenty-four inches, is sometimes a perennial, and lives in sour soil. But do not lose any sleep, dear reader, if you fail to distinguish the two weeds. If the weed is in a pasture or meadow, simply call it the Meadow barley, and if you see it along the fence rows or waysides still call it the Meadow barley, or if you want to impress your hearers with your great learning call it *Hordeum nodosum.*

Hordeum is the ancient name for barley. *Nodosum* refers to the node, not of the stems but of the heads. The heads, you will remember, separate at their nodes.

TECHNICAL DESCRIPTION

See Squirrel-tail Grass for technical description of *Hordeum* (Tourn.) L.
Hordeum nodosum L. Spike 2–8 cm. long, about 1.5 cm. wide; *all the glumes awn-like,* 1–1.5 mm. long.—Thin dry soils, Indiana, Minnesota, and northwest, south to Tennessee and Texas. (Eurasia.)

YELLOW NUT GRASS

[*Cyperus esculentus* L.]

AND IT is not a grass at all. It is a sedge, this so-called Yellow nut grass. Sedges may look like grasses to the casual observer, but aside from their blades for leaves, sedges and grasses are not very much alike. Grasses have round stems (culms) with nodes along them. The leaves, or blades, arise from the nodes and sheath the stem for a distance above each node. The sedge stems are usually three-cornered and without nodes. At the top of a sedge stem there is usually a cluster of flower-bearing branches that arise from the center of a whorl of small blades. The flower spikes of this particular sedge are so arranged on the branches that bear them that they look like little fans. Three or four little yellow fans at the top of a three-cornered stem assure one that he has found the Yellow nut grass or some other sedge that very much resembles it.

The nuts that give the plant its name are little tubers on its roots. They are usually few and small, but if, when the plant is pulled up, some little bulblike tubers are found attached to the ends of some of the roots, and if these little tubers taste like almonds or filberts, the plant is the Yellow nut grass.

Like most of the sedges this one is partial to wet places. It enjoys paddling in the mud if not in the water. One of the best ways to eradicate it is to drain the spot it has seized upon. None of the lawn grasses will tolerate wet feet. They soon sicken and die if the ground is too wet, but that is exactly what the sedges crave. Some of the sedge species might be called the ducks of the plant kingdom, but this one is a sort of amphibian. It likes to play in the water but it also rejoices in the sunshine. It usually takes places where the grasses cannot grow, but after it is once in, even if the spot is drained, it is difficult to eradicate because of its tubers. It often requires a year or more of clean cultivation in well-drained places to rid the ground of the pest.

It is a pest but it is a plant with possibilities. It is said that nut sedge tubers are sold on the Italian market. They could be sold on any market, for through search and selection it ought to be possible to find a nut-sedge plant that would produce many large tubers. If such a plant could be found it would be worthy of cultivation, and another item would thus be added to our vegetarian diet.

The plant has several names even though it is native to this country. It is called Nut sedge (the most sensible name it has), Cufa, Northern nut grass, Coco, Coco nut, Rush nut, Edible galingale, Earth almond, and Ground almond. The generic name, *Cyperus* (pronounced Sī-pē′rus), is new Latin from the Greek and means a sedge. The specific name, *esculentus,* means edible. Therefore *Cyperus esculentus* L. means the edible nut sedge named by Linnæus.

TECHNICAL DESCRIPTION

Cyperus (Tourn.) L. Spikelets many–few-flowered, mostly flat, variously arranged, mostly in clusters or heads, which are commonly disposed in a simple or compound terminal umbel. Scales 2-ranked (their decurrent base often forming margins or wings to the hollow of the joint of the axis next below), deciduous when old. Stamens 1-3. Style 2-3-cleft, deciduous. Achene lenticular or triangular, naked at the apex.— Culms mostly triangular, simple, leafy at base, and with one or more leaves at the summit, forming an involucre to the umbel or head. Peduncles or rays unequal, sheathed at base. All flowering in late summer or autumn. (The ancient name.)

Cyperus esculentus L. Similar; culms (3–9 dm. high) equaling the

Fig. 16. The Yellow Nut Grass is a sedge

leaves; umbel often compound, 4–7–rayed, much shorter than the long involucre; *spikelets numerous, light chestnut or straw-color, acutish,* 0.5–1.5 cm. long; *scales ovate or ovate-oblong, narrowly scarious-margined, nerved,* the acutish *tips rather loose;* achene oblong-obovoid. —Low grounds, along rivers, etc.; spreading extensively by its small nut-like tubers and sometimes becoming a pest in cultivated grounds. (Eurasia.)

CALAMUS

[*Acorus calamus* L.]

IF THE Calamus did not require wet feet it would be a bad weed. As it is, it is the one weed that man plants where it can have its own way. Few are the country boys who have not tasted the hot, sweet root-stalks of calamus, and so when those boys become men and owners of farms that have boggy spots on them the taste of calamus is remembered and a root or two is set in the marshy place. Fortunately the plant will not leave the wet place, but in time it will fill every inch of it. Thus many a wet area that might grow edible grass grows—all because of a memory— the inedible calamus. That is why little swales are often filled with the weed, and also why it is not in other swales: it has not been planted there.

Calamus is an interesting plant. It looks like iris to the casual observer. Its blades are much longer and narrower than are those of most iris plants, but like the iris it is called a flag. Sweet flag, Sweet sedge, Sweet root, and Sweet rush are its common names. Calamus, however, its botanical name, is its most used appellation. This is likely because the plant was considered medicinal in days gone by and physicians who knew the botanical name prescribed the chewing of calamus root for stomach troubles, especially for flatulence. It is said to be actually good for flatulence, and it is better than the much-advertised remedies for halitosis. These are about the only virtues the weed has, except that it will always stay in what for most plants is a worthless environment.

Many who know the plant have never seen its flowers or seed.

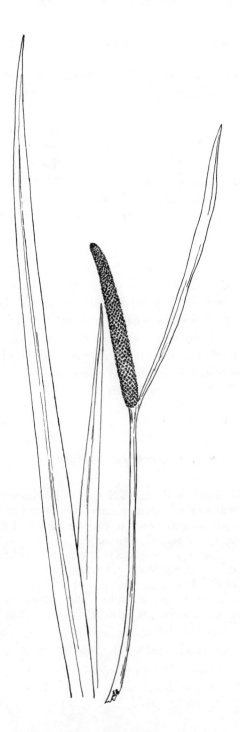

FIG. 17. Calamus or Sweet Flag

The reason for this is that it never flowers except where the marsh is wet enough to keep the roots immersed in water the greater part of the year. It will grow where there is much less water than that, but in the dryer places its reproduction is entirely by way of the root-stalk. The flowers and seed are on what is called a spadix. The spadix has a tender core and a taste that is much milder than that of the root-stalk. For this reason children eat the spadices, flowers and all, and declare them to be toothsome. They are not unpleasant to the taste nor is there any danger in the eating of them.

Calamus belongs to the arum family, the same family that contains the Calla, the Jack-in-the-pulpit, the Green dragon, and Skunk cabbage. The family is known by the peculiar way the flowers are borne. They are thickly set in a spike which has a fleshy axis. Such a spike is called a spadix.

The complete botanical name of calamus is *Acorus calamus* L. *Acorus* is likely the ancient Greek name of the plant or of some other plant in the Arum family. *Calamus* is the Latin name for reed. Translating the name, then, we have "the reedy arum named by Linnæus."

TECHNICAL DESCRIPTION

Acorus L. Sepals 6, concave. Stamens 6; filaments linear; anthers kidney-shaped, 1-celled, opening across. Ovary 2-3 celled, with several pendulous orthotropous ovules in each cell. Fruit at length dry, gelatinous inside, 1-few seeded.—Aromatic, especially the thick creeping rootstocks (*calamus* of the shops). Leaves sword-like; the upper and more foliaceous prolongation of the scape may be considered as a kind of open spathe. (*Acorus,* the ancient name, of no known meaning.)

Acorus calamus L. Scape leaflike and prolonged far beyond the (yellowish-green) spadix.—Margins of rivulets, swamps, etc. (Eurasia.)

SPIDERWORT

[Tradescantia reflexa L.]

THE MASSES of beautiful blue flowers seen along railways and roadsides in late springtime are almost sure to be made up of one or more of the spiderworts. The spiderworts are not bad weeds. In truth they are so innoxious that they might well be left out of a weed book were it not for the fact that they are so common and so conspicuous when in bloom that nearly every one who sees them exclaims, "What is that blue flower?" That is why a spiderwort is treated here: to inform the unknowing that that blue flower is just a weed.

This species is probably as widely spread and as often seen as any of the seven species in the genus, and it has at least one weedy character: it is a perennial, and when it is once established it is capable of holding its position against odds. It is a true weed in this respect. It has no special method of seed dispersal and so has to rely on the wind and the resiliency of its lithe stems. For this reason a single plant becomes the center of a colony of plants, and the rushing trains scatter the seeds up and down the tracks. This species (*Tradescantia reflexa* L.) seems partial to the graded fills and the undisturbed land of the railroad right of ways, especially where the ground is wet.

The spiderworts are beautiful enough to be used in flower gardens. They serve exceptionally well in rock gardens and on terraces where cultivation is not possible. None of the species thrives under cultivation, and he who wants to use one of them in a decorative way should remember this fact. "Give it a place and leave it alone," is the slogan by which to grow a wild spiderwort. In a rock garden a mass of spiderwort makes a pleasing contrast with a mass of wild columbine, since they bloom at the same time, and since both plants give a wild, natural air to the place.

The name spiderwort means spider plant, and refers to the hairy stamens which look like the hairy legs of a spider. The name of the genus is in honor of one Tradescant who was the

FIG. 18. The Spiderwort. Note the hairy stamens
at the center of the flower

gardener to Charles the First of England. The specific name, *reflexa,* refers to the reflexed peduncles of the flowers before and after blooming.

TECHNICAL DESCRIPTION

Tradescantia (Rupp.) L. Flowers regular. Sepals herbaceous. Petals all alike, ovate, sessile. Stamens all fertile; filaments bearded. Capsule 2–3–celled, the cells 1–2–seeded.—Perennials. Stems mucilaginous, mostly upright, nearly simple, leafy. Leaves keeled. Flowers ephemeral, in umbeled clusters, axillary and terminal, produced through the summer; floral leaves nearly like the others. (Named for the elder *Tradescant,* gardener to Charles the First of England.)

Tradescantia reflexa Raf. *Slender,* glabrous or nearly so, *glaucous;* leaves narrow, linear-attenuate from a lanceolate base, strongly involute; umbels terminal on the stems and branches, many-flowered; *narrow bracts and glabrous pedicels soon deflexed; sepals* ovate-lanceolate, 8–13 mm. long, *glabrous except at the often tufted tip;* petals blue, 10–14 mm. long.—Wet places, Ohio to Michigan, Minnesota, Kansas, Texas, and South Carolina.

WIRE GRASS

[*Juncus tenuis* Willd.]

EVERY country boy or girl who has walked along a well-beaten path on a wet morning and has found his or her bare ankles bespattered with tiny gelatinous seeds has made the acquaintance of the weed of this sketch. No frequently used path in the United States and Canada is likely to escape a margin of Wire grass, Path rush, Field rush, Slender yard rush, or Poverty grass, its other names.

It is interesting to note that so many of its names have *rush* attached to them. Rush is the Anglo-Saxon name for the whole family to which this little path runner belongs. The name is as old as the early English, the middle French, and the middle high German. It probably refers to the sort of stem these plants have, but it might refer to their use. Rushes were used by the

FIG. 19. Wire Grass or Juncus

early English for many things: to strew on floors, to make into mats, bands and ropes, and some of them—perhaps none of this family but some plants called rushes, at least—were used as light wicks.

The generic name, *Juncus,* is from the Latin word *jungere,* meaning to join, and so refers to the part played by the rushes in band and rope making. This particular species, *tenuis,* was probably too small to be of much worth, but there are species in the family with stems three feet long, and it was these that gave the names rush and Juncus to the group.

The plant has little reason for being treated in a book of weeds, but the botanist dislikes to hear a plant called what it is not, and the teacher always yearns to teach the ignorant. The weed is **not** a grass any more than a smartweed is a clover. It is a juncus, a rush, but not a grass. So when you walk along a path again, just remember that the wiry stems which brush your ankles are not of a grass at all, but are actually juncus or rush stems, and you will be much wiser in the eyes of the botanist.

TECHNICAL DESCRIPTION

Juncus (Tourn.) L. Capsule 3–celled, or 1–celled by the placentæ not reaching the axis. Stamens when 3 opposite the 3 sepals.—Chiefly perennials, and in wet soil or water, with pithy or hollow and simple (rarely branching) stems, and cymose or clustered small (greenish or brownish) flowers, chiefly in summer. (The classical name, from *jungere,* to join, alluding to the use of the stems for bands.)

Juncus tenuis Willd. Stem wiry (0.5–6 dm. high); *cyme* 1–8 cm. long, loose, or barely crowded; *flowers green* (3–4.5 mm. long), *mostly aggregated at the tips of the branches;* sepals lanceolate), very acute, spreading in fruit, longer than the ovoid retuse scarcely pointed green, falsely 1–celled capsule; anthers much shorter than the filaments; style very short; seeds small (3–4 mm. long), delicately ribbed and cross-lined.—Fields and roadsides, very common. June–September. (Europe, North Africa.)

WILD GARLIC

[*Allium vineale* L.]

ONLY those who have smelled the breath of cows pasturing on wild garlic, or have tried to drink the milk of those cows, can fully appreciate the stinking importance of this outlaw of the lily family. No weed has a worse reputation with dairy farmers, and it is one of the worst weeds in wheat fields. It is true that the garlic bulblets can be made to shrivel and so be removed from the wheat several days or weeks after it is threshed, but wheat containing the bulblets is always "docked" and until the garlic is removed it is unfit for milling.

Why a plant was created with an odor so vile as that of the garlic is a question that requires a course in evolution to answer and to understand. Natural selection might have had something to do with it. Perhaps it has been the "stinkingest" bunches that have survived in days gone by, but if this is true, then there has been developed along with it a host of pasturing animals with more and more defective olfactories. Whether the plant is exceedingly nutritious and the cows and sheep eat it as man eats onions in spite of the smell, or whether it is a delectable flavor and fragrance the cow experiences when she mows down great swaths of the stinking weed no one but the cow will ever know, but we do know that her voraciousness spoils her breath, her milk and her meat, and yet she gets fat and exceedingly contented when she eats the accursed plant.

Much has been said on the eradication and control of wild garlic. Big bulletins on the subject have been written and sent out from Washington and from state experiment stations, but still the garlic spreads. It is in pastures and lawns, in cultivated fields and gardens, it is in hay and wheat, in milk and meat, and it is getting worse all the time. And it has been doing so since it came to this country away back in sixteen hundred and something. The march has been a slow one from east to west, but a steady one, and every farm entered has been conquered and held. The farmer almost never notices the invasion until it is too late. He does not

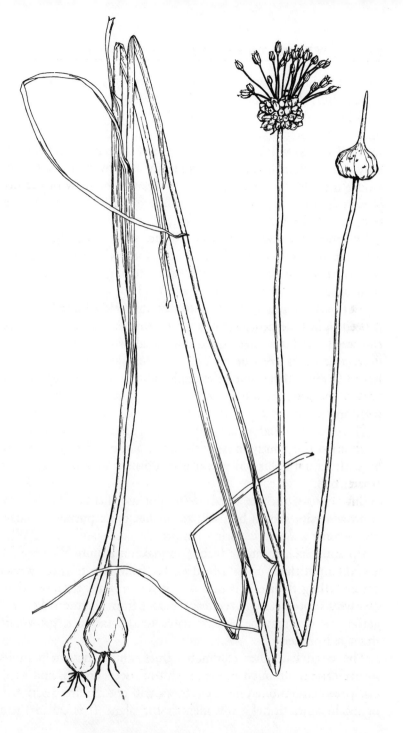

FIG. 20. Wild Garlic, It is one of the worst

discover it till the cows come home some evening and their breath knocks him down, metaphorically speaking. Then if he is the man he should be he gets up and tries to do something about it, but, as was stated above, it is too late. From that time on he takes his dockage on wheat and learns the trick of pasturing his cows only three or four hours a day. He learns that by so doing he can reduce the odor to a point where the cream-sampler whose senses of taste and smell have become dulled to garlic cannot detect a trace of it. At any rate he sells his cream.

Yes, there are several ways to kill the weed. It can be sprayed with oil, iron sulphate, sulphuric acid and sodium chlorate, but the very best way to kill it is to turn in the hogs without rings in their noses, or to plow and replow a field for a year, or until no sign of garlic appears, and then to sow the field to alfalfa, if hay is needed, or to lespedeza if it is pasture that is wanted. Of course the weed has little effect upon cultivated crops and so may be disregarded in a corn or cotton field, but please remember that it will be right there when the field goes into oats, wheat, clover, hay, or pasture. In other words garlic is no respecter of crop rotation.

There is no weed worse than wild garlic and its complete eradication is devoutly to be wished, but garlic and sin will likely be with us until this old world goes up in smoke or becomes a frozen ball.

But why not use the weed? Why not use this bit of adversity as all adversity should be used: to further one's purpose? Garlic can become a valuable spring pasture for cattle and sheep. Why not pasture such meat-producing animals on it until they are fat and then put them up in a feed lot for from two to three weeks before selling them? It is said that it requires a little more than two weeks completely to remove the odor from the meat. Perhaps garlic pastures may become valuable to the fat-stock grower if managed properly.

The weed has other common names which may help in its identification, although any one who has seen onions, and especially seed onions or a bunch of chives, will not require descriptive or local names to help him identify the plant. It is called Field garlic, Crow garlic, and Wild onion. The botanical name, *Allium vineale,* means the garlic that grows in vineyards. The generic name, *Allium,* is the ancient name for garlic, and *vineale* means pertaining to vines or vineyards.

TECHNICAL DESCRIPTION

Allium (Tourn.) L. Perianth of 6 entirely colored sepals, which are distinct, or united at the very base, 1-nerved, often becoming dry and scarious and more or less persistent; the 6 filaments awl-shaped or dilated at base. Style persistent, thread-like; stigma simple or only slightly 3-lobed. Capsule lobed, loculicidal, 3-valved, with 1-2 ovoid-kidney-shaped amphitropous or campylotropous black seeds in each cell.—Strong-scented and pungent herbs; the leaves and usually scapose stem from a coated bulb; flowers in a simple umbel, some or all of them frequently replaced by bulblets; spathe scarious, 1-2-valved. (The ancient Latin name of the Garlic.)

Allium vineale L. Stem slender (3-9 dm. high), clothed with the sheathing bases of the leaves below the middle; *leaves terete and hollow,* slender, channeled above; *umbel often densely bulbiferous; filaments much dilated, the alternate ones cuspidate* on each side of the anther.—Moist meadows and fields, locally abundant, Massachusetts to Missouri, and Virginia. June. (Naturalized from Europe.)

III

WEEDS THAT ARE NOT GRASSLIKE

Fig. 21. Yellow Dock in bloom

YELLOW DOCK

[*Rumex crispus* L.]

ONE OF the most persistent of the perennial weeds is the Yellow dock, or, as it is called in different localities, Sour dock, Curly dock, and Narrow-leaved dock. It has a long tap-root that seems to be able to stand any amount of punishment if only a part of it is permitted to stay in the ground. Allow the plant to stand a year or two, and so give it a chance to drive that food-filled root deep into the earth, and nothing but a complete extraction or a thorough salting will stop dock plants from appearing in that place. Spudding a big yellow dock plant does little more than prevent the seed production of the current year. Not just one but several little docks will arise, Phœnixlike, from the fires burning that store of fuel in the root left below.

And yet, Yellow dock is not a bad weed. It does get into cultivated fields occasionally, but it is a lazy farmer who cannot keep it out of such places. It is more often seen in unmolested spots than anywhere else; along fence rows and waysides and in swales of old meadows and pastures.

Strange as it may seem, the Yellow dock has medicinal properties enough to place the plant in the United States Pharmacopœia in the days before the gay nineties. The gay nineties removed it, but it was in there until then. The root has long been used in home remedies. It can be made into tonics, laxatives, and salves. It is mildly astringent, and it is said that its root makes a good dentifrice if it is dried and ground to a powder. The leaves of Yellow dock are widely used as greens—pot-herb dishes. They are savory and take the place of Poke leaves when the mixture is made up principally of mustard and dandelion. The rather loose rosettes of long tender leaves are selected for greens. After the stalk appears the leaves become rather tough.

The tall, stiff stalk with its open panicle-like top made up of spikes crowded with three-angled fruits is enough to identify the weed in the latter part of the summer. The long, lance-shaped leaves with their wavy or "crisped" edges should identify it in the early part of the season. The "Yellow" part of the name refers to the panicled top in its blooming stage. It is not quite yellow, but yellowish when seen at a distance.

There is another dock with which the yellow dock may be confused and that is the one called "Broad-leaved." The fruiting tops of the two species are somewhat alike, but their leaves are very different. The Broad-leaved species has very broad leaves, from the midst of which the stalk arises, and the veins of these leaves are red or reddish in color. The yellow dock leaves are never broad in comparison with their length, their veins are never red, and their edges are never smooth. It is called curly dock because of the curled edges of its leaves. The botanical name, *crispus,* means curly.

Rumex is the Latin name for the docks, and *dock* is the Anglo-Saxon *docce,* designating this particular plant and all of its relatives. So here is one plant with a common name that is almost if not quite as weighted with antiquity as is its scientific name.

To reiterate, Yellow dock is not a bad weed, but it often becomes obnoxious enough to need eradication. When such is the case the best way to achieve effective eradication is to pull up the weeds. When the ground is filled with water this can be done if the plants are not too old. If they are too old and their roots are too big, chop off the tops of the roots an inch or more below the surface of the ground and pile a double handful of salt on top of each of the newly made wounds. Roots so treated will never have to be pulled.

TECHNICAL DESCRIPTION

See Sheep Sorrel for the technical description of the Genus, *Rumex* L. *Rumex crispus* L. Smooth, 0.9–1.6 m. high; *leaves with strongly wavy-curled margins, lanceolate,* acute, the lower truncate or scarcely heart-shaped at base; *whorls crowded in prolonged wand-like racemes, leafless above;* pedicels with tumid joints; *valves round-heart-shaped, obscurely denticulate* or entire, 4–6 mm. broad, mostly all grain-bearing; the *grains very plump, subglobose to ellipsoid, with rounded ends.* —In cultivated and waste ground, very common. (Naturalized from Europe.)

SHEEP SORREL
[*Rumex acetosella* L.]

SHEEP SORREL is a communist. It waves the red flag wherever it moves in, and it moves in wherever it finds the democratic grasses struggling against adverse conditions. Small though it is, its snake-like rootstalks crawl under and among the grass roots and send up new "reds" among the grass bunches. A patch of Sheep sorrel never gets any smaller, and nothing but healthy soil conditions will keep it from colonizing wherever a whirlwind lets fall one or more of its seeds. Its seeds ride in on whirlwinds. It is truly a child of calamity.

If the farmer does not know the weed as Sheep sorrel (it is the farmer who needs to know this weed) he may know it as Red sorrel, Field sorrel, Sour weed, Red-topped sorrel, Cow sorrel, Sour leek, Horse sorrel, Gentlemen's sorrel, Toad sorrel, or just Sorrel. The word sorrel is used for any sour-tasting plant, and has been so used for hundreds of years. It is derived from both the old French and the old high German, and it has always been applied to this weed and to some of its cousins as well as to the sour-tasting oxalis species.

The botanical name is *Rumex acetosella* L. The word *Rumex* is from the Latin and means Sorrel. *Acetosella* is from the Latin also, and means "the little vinegar plant." There is another Rumex called *Rumex acetosa*. It looks and tastes very much like the Sheep sorrel but it is much larger; so this one, the Sheep sorrel, has a name with the diminutive ending. It is the little *acetosa* or *acetosella*.

The weed is so easily identified that a description along with the picture is superfluous. Suffice it to say that when one finds in his field or lawn a little plant with halberd-shaped leaves that taste like vinegar he has surely found the Sheep sorrel. And he can be almost as certain that that particular spot of soil needs attention. Either the ground has become too sour—the usual thing

FIG. 22. Sheep Sorrel, a plant communist

—or the grass has exhausted its food materials. Contrary to the common belief Sheep sorrel does not grow best in sour soil, but it is a good indicator of sour soil, for it can and does grow in soil too sour for other plants, especially grasses. Nearly all grasses require a comparatively sweet soil, and Sheep sorrel is not in the least particular about such things. It is full of vinegar anyway, so why worry about a little acid in the soil?

The weed is so small, never more than a foot high, that the best way to keep it out of a lawn or meadow is to keep the sod vigorous. When it becomes established it means that the sod is in need of a thorough overhauling. It should be plowed up, whether it is meadow or lawn, put into shape and reseeded. This is the very best way to fight our little red communist, the Sheep sorrel.

TECHNICAL DESCRIPTION

Rumex L. Calyx of 6 sepals; the 3 outer herbaceous, sometimes united at base, spreading in fruit; the 3 inner larger, somewhat colored (in fruit called *valves*) and convergent over the 3-angled achene, veiny, often bearing a grain-like tubercle on the back. Stamens 6. Styles 3; stigmas tufted.—Coarse herbs, with small and homely (mostly green) flowers, which are crowded and commonly whorled in panicled racemes; the petioles somewhat sheathing at base. (The ancient Latin name; of unknown etymology.)

Rumex acetosella L. Low (1-3 dm. high); leaves narrow-lanceolate or linear, halberd-form, at least the lowermost, the narrow lobes entire, widely spreading; *pedicels jointed at the summit; sepals scarcely enlarged in fruit, exceeded by the naked achene.*—A common weed. (Naturalized from Europe.)

KNOTGRASS

[*Polygonum aviculare* L.]

To BEGIN with, Knotgrass is not a grass. It is a member of the buckwheat family and a cousin of the smartweeds. It probably gets the grass part of its name from the fact that it often takes the place of grass where poultry and children, by much tramping, destroy the grassy sod. The weed is especially fond of well-packed soil, and for this reason one often finds it bordering or even growing in well-beaten paths.

Like most of the weeds that came from the Old World this one has many English names: Ninety-knot, Centinode, Nine-joints, Allseed, Bird's-tongue, Swynel grass, Swine's grass, Red-robin, Armstrong, Cow-grass, Hog-weed, and Pig rush. But Knot-grass is the favorite name and has been so for many years. Shakespeare used it in *Midsummer Night's Dream,* calling it "the hindering knotgrass" in keeping with the belief of the time that the juice of this weed would interfere with the growth of children and young farm animals. Those were the days when weeds as well as brains were potent. The same thing that connected bad weeds with witches sent Drake off on his voyages and created unsurpassed plays in the mind of Shakespeare.

Knotgrass is often confused with one of the native lespedezas, a small legume. There is only one reason for such confusion and that is lack of observation. The lespedezas have three-parted leaves: that is, there are three leaflets on each leaf stem. The knot-grass has *simple* leaves. It is true that the blade of its leaf is about the size of each part of a lespedeza leaf, but the knotgrass leaf is not only a simple leaf, but has, also, a very short stem that has at the base of it a *stipule* which surrounds the stem of the plant above the node where the leaf is attached. This is the mark of the buckwheat family: many joints with stipules surrounding the stem above the joints. The botanical name, *Polygonum,* is from the Greek and means "many knees"—many joints.

There are many species of knotgrass but they are all very much alike in appearance and habits. The one here treated is the one most often seen, and the one that has medicinal properties worth knowing. It is said that the juice of this weed squirted into the nose will stop nose-bleed. Its tea is useful in cases of diarrhea and bleeding piles. It can be used for all sorts of hemorrhages.

The weed is relished by cows and poultry, and its seeds are eaten by small birds. The specific name, *aviculare,* is from the

Fig. 23. Knotgrass, a weed of the misused lawn

Latin and means small birds. The seeds are produced at nearly every node and in few-seeded clusters. The inconspicuous flowers that appear before the seed are worth seeing. Sometimes their tiny perianths are red enough to warrant the name "Red Robin." They are usually merely pink-margined, however.

TECHNICAL DESCRIPTION

Polygonum (Tourn.) L. Calyx 4-6 (mostly 5)-parted; the divisions often petal-like, all erect in fruit, withering or persistent. Stamens 3-9. Styles or stigmas 2 or 3; achene accordingly lenticular or 3-angular. . . . Pedicels jointed.—Ours all herbaceous, with fibrous roots, flowering through late summer and early autumn. (Name from two Greek words meaning *many* and *knee,* from the numerous joints.)

Polygonum aviculare L. Slender, *mostly prostrate or ascending, bluish-green;* leaves lanceolate, 6–20 mm. long, usually acute or acutish; *sepals hardly* 2 mm. long, green with pinkish margins; stamens 8 (rarely 5); achene dull and minutely granular-striate, mostly included.—Common everywhere in yards, waste places, etc. (Eurasia.)

SHOESTRING SMARTWEED

[*Polygonum Muhlenbergii* Wats.]

THE FAVORITE haunt of the Shoestring smartweed is the heavy black soil of bottom lands. It is sometimes called swamp smartweed and Swamp persicaria but these names should be applied to another species that actually does live in swamps: *Polygonum amphibium.* The subject of this sketch can and does live and thrive in comparatively dry soils, but it is at its best in heavy, wet places; in gumbo and wet, heavy clays. It is sometimes called gumbo shoestring; also devil's shoestring.

The "shoestring" part of the name refers to the perennial root-stalks, which do resemble black shoestrings, especially after they have been plowed out and allowed to dry. Like all perennial plants

FIG. 24. The Shoestring Smartweed and
two developing shoestrings

with underground stems the Shoestring smartweed is scattered and planted by the plow and harrow, and because of this in low, wet grounds it plays a close second to Man-under-ground among the weeds of such places.

One of the striking characters of the weed is its refusal to bloom profusely as do the other smartweeds. It depends far more upon its rootstalks for propagation than upon its seeds. One is likely to see acres of the weed standing in a mass of from twenty-four to thirty inches high without a flower showing. The plants that do bloom are usually scattered apart from the others, and even these produce very short spikes of a few rose-colored flowers, similar in shape to those of the Lady's thumb and the Pennsylvania smart-weed.

The best way to prevent the transplanting of the Shoestring smartweed is to use both the disc breaking or turning plow and the disc harrow, commonly called the disc. It is possible completely to eradicate the weed with these implements. In some places where it is very bad the disc cultivator should be used also.

It is difficult to employ the weed as a fertilizer because of its shoestring rootstalks, but it can be turned under and made to serve as a green manure provided the young shoots from the root-stalks are cut down with a disc every time they appear. This, by the way, is one of the best methods to use in eradicating the weed.

The botanical name of the plant is *Polygonum Muhlenbergii* Wats. This means the many-kneed plant named for the botanist, Muhlenberg, by the botanist Sereno Watson. The generic name, *Polygonum,* means "many knees" and is very descriptive of most of the species, but less so of this one.

TECHNICAL DESCRIPTION

See Knotgrass for the technical description of *Polygonum* (Tourn.) L. *Polygonum Muhlenbergii* (Meisn.) Wats. Perennial, in muddy or dry places, rarely in shallow water, decumbent or suberect, scabrous with short appressed hairs; *leaves lanceolate to ovate, narrowly acumi-nate* (1–2 dm. long); *peduncles hispid and often glandular; spikes* 3–10 cm. long, often in pairs; flowers and fruit nearly as in the last (this means bright rose-colored flowers)—Quebec and Maine to Florida, and westward.—Exceedingly variable in foliage and pubescence; aquatic states often have essentially glabrous and cordate leaves, while in plants of drier situations these are sometimes narrowly lanceolate, acute at base, and conspicuously appressed-pubescent on both surfaces.

PENNSYLVANIA SMARTWEED

[*Polygonum pennsylvanicum* L.]

THIS IS one of the smartweeds without the smart. It has the generic appearance, however, and has far more reason for being called a smartweed than the butterfly has for being called a butterfly, but just the same there would be less strain on the truth if instead of "Smartweed" it were known by some of the other names it has acquired, such as Purple top, Hearts-ease, Glandular persicary, Swamp persicary, etc.

This species is often confused with a near cousin known as Lady's-thumb, Persicary, Spotted smartweed, Heart-weed, Oxheart, Spotted knotweed, Red Shanks, Lover's pride, etc. The two weeds are found together, and so far as the husbandman is concerned they are the same thing, and *Polygonum pennsylvanicum* or *Polygonum persicaria* are all one or all nothing to him, but there is satisfaction in seeing and knowing the difference. The Pennsylvania smartweed, *Polygonum pennsylvanicum* L., is more erect than is *Polygonum persicaria*. Its leaves are immaculate, that is to say unspotted, and its fused stipules (*Ocreæ*, pronounced ŏk'rē-ē, singular *ocrea* ŏk'rē-à) are without bristles along their margins. The other, the Lady's-thumb, has the mark of the lady's dirty thumb on every leaf, and its ocreæ have at least a suggestion of bristles around their margins. And Lady's-thumb is bigger, more branched, and is inclined to lie down more than is the Pennsylvania species. Of course the obvious conclusion is that the Pennsylvania species got its name from the State of Pennsylvania, but its immaculate leaves and erect stems suggest the Quaker, and perhaps Linnæus, who was something of a humorist, had this in mind when he named the weed.

Fig. 25. The Pennsylvania Smartweed
is a valuable soil builder

This plant, along with Lady's-thumb, should be used in soil building. Buckwheat is sometimes used for that purpose, and these weeds, belonging as they do to the buckwheat family, serve even better than cultivated buckwheat does. They will make a better growth on poor soil than buckwheat will and so will put more fertilizer into the ground than the buckwheat can. Such weeds as these should be welcomed by the farmer and gardener, but not unless they are to be used. From the time the seeds sprout until the plants are in full bloom they are capable of enriching the soil they grow in if they are but plowed or hoed into it. He who fails to use these two valuable but often troublesome weeds as soil builders has no appreciation of the need of decaying vegetable matter in soils.

The name, *Polygonum,* means many knees. Many-knees from Pennsylvania is a free translation of the botanical name *Polygonum pennsylvanicum* L.

TECHNICAL DESCRIPTION

See Knotgrass for the technical description of *Polygonum* (Tourn.) L.

Polygonum pennsylvanicum L. *Annual; leaves lanceolate; branches above and especially the peduncles beset with stipitate glands; flowers uniform,* bright rose-color, in short erect spikes, often on exserted pedicels; stamens usually 8; achene nearly orbicular, over 2 mm. broad, at least one surface concave.—Moist soil, in open waste places, central Maine, westward and southward.—Neither the stamens nor style conspicuously exserted.

SMARTWEED

[*Polygonum hydropiper* L.]

PŎ-LĬG'Ō-NŬM HY-DRŎ'PĬ-PĔR is certainly a mouth-filling name, and he who tastes the weed gets a mouthful, too. This is the one and only Smartweed, Water pepper, Pepper plant, Biting persicaria and Biting knotweed. It is also called Red knees and Red shanks.

There are a great many species of Polygonum, several of which look somewhat like this weed, but most of them lack the "smart" taste entirely and none of them possesses the fire of hydropiper. One has to be careful even in handling this weed, for if the fingers that grasp it come in contact with the eyes there will be a stinging sensation that will cause the careless one to respect this little plant vixen as long as he lives.

It is not strange that a plant with so much of seeming potency in its juices should be considered as at least one of the answers to a sick man's prayer. It was at one time used in many ways for many ills, but time has proved it to be practically worthless. Its fire resembles that of the mustards, but the mustards have something that the smartweed does not have.

The plant loves wet places. Water pepper is a good name for it. That is the meaning of its specific name, *hydropiper*. It likes wet places, but it can and does grow in ground that is dry enough to be cultivated. It may become a weed in a good garden and it is often seen in chicken lots where dog-fennel and another polygonum, a distant cousin of the smartweed known as knotweed, abound. It is often found in wet places in pastures, since grazing cattle soon learn to leave it alone. A hungry calf may be tempted to take a bite of smartweed, but if one has any sympathy for dumb animals he will never tempt another calf, and it will be impossible for him to tempt the same calf a second time. Tom Sawyer's Peter cut no more shines after his dose of pain killer than a calf will cut after it has been fed a mouthful of smartweeds. The reputation of this little weed—and it is a little weed, never becoming more than two feet high, and usually not more than half that height—has branded all of the polygonums that resemble it with the name of "Smartweed." Lady's-thumb and the Pennsylvania smartweed are two examples of this. They are both called smartweeds in spite of the fact that there is no "smart" in them.

The smartweed can easily be identified by its pungent taste, but its visible characters are enough to distinguish it from other polygonums. The sheaths around the stems at the base of each leaf are a polygonum character. All plants having that are related to the

Fig. 26. The Smartweed with its
bristle-edged ocreæ

smartweed, and are all in the great buckwheat family. The smart-weed has these fused sheaths, "stipules," as they are called, and they are a little different from those of the other species of the genus in that they have bristly margins. The smartweed has the nar-rowest, sharpest-pointed leaves of the lot, its flowers are small and greenish-white and are in terminal, nodding spikes. There are also a few short, lateral spikes in the axes of the upper leaves.

If one will keep these facts in mind while learning the polyg-onums he can be rather confident of his decision on this weed. If he is in doubt, however, all he has to do is taste a leaf.

TECHNICAL DESCRIPTION

See Knotgrass for the technical description of *Polygonum* (Tourn.) L. *Polygonum hydropiper* L. Annual, 3–6 dm. high, smooth; *leaves* narrowly lanceolate, *very acrid and peppery; spikes nodding,* usually short or interrupted; flowers mostly greenish; *stamens* 6; style 2–3-parted; *achene dull,* minutely striate.—Moist or wet grounds; appar-ently introduced southeastward, but indigenous northward and west-ward. (Europe.)

LAMB'S QUARTERS

[*Chenopodium album* L.]

LAMB'S QUARTERS, Pigweed, White chenopodium, or *Chenopo-dium album,* as the botanist calls it, is one of the most common of the big garden weeds in the Mississippi Valley. It is an easy weed to recognize and a fairly easy one to control, but it delights in the lazy owner of the rich garden patch.

The botanical name means the white goose foot. Goose-foot is a name given the family to which this weed and the cultivated beets belong. The man who named the family *Chenopodiaceæ* saw that the leaves of many of its members resembled the foot of a goose and in the white chenopodium the resemblance is as strik-ing as in **any of** the species.

FIG. 27. Lamb's Quarters at the stage
when few recognize it

The white (*album*) of the name refers to the silver sheen that the leaves have. They are not very silvery but they do have a sheen and they are whiter than the leaves of most other weeds, and so the plant deserves this name, *album*.

Why the weed was ever called Lamb's quarters is not so easily explained. The young plants make very tender greens, and so it may have been that some one eating a dish of white Chenopodium greens let his imagination get the better of him and fancied that these tenderest of all greens were like the tenderest of all meats and called them Lamb's quarters. Or it may have been the fancied resemblance of the shape of the leaf to that of a leg of lamb. Whatever it was, the weeds seem to be known when known at all either as Lamb's quarters or as Pigweeds. They are relished by swine and so they deserve to be called Pigweeds.

The weed can be easily recognized by its silvery sheen, its straight central stem which branches when the plant is not crowded but which has few branches when it grows as it likes to grow, crowded close among its fellows; and by the loose panicles of tiny ball-like flowers and seeds that are as green as the rest of the plant. After the seeds have been formed and before they are ripe the leaves and seed-balls sometimes turn red. Plants may be seen in late summer red enough to attract attention and to cause even casual observers to ask, "What is that red weed?" It is one of the last weeds to be killed by the frosts of autumn and it is one of the best indicators of good soil. If plowed under soon enough it makes a good fertilizer, but its stem becomes hard even before the flowers bloom, and so if it is used as a fertilizer it should be under the soil before the panicled top forms. Learn to know the weed in the early stages if you care to make food of it either for yourself or for the bacteria of your soil.

TECHNICAL DESCRIPTION

Chenopodium (Tourn.) L. Flowers all bractless. Calyx 5 (rarely 4)-parted or –lobed, more or less enveloping the fruit. Stamens mostly 5; filaments filiform. Styles 2, rarely 3. Seed lenticular, horizontal (*i.e.*, with its greatest diameter at right angles to the floral axis) or vertical;

embryo coiled partly or fully round the mealy albumen.—Weeds, usually with a white mealiness, or glandular. Flowers sessile in small clusters collected in spiked panicles. (Named from two Greek words meaning a goose and foot, in allusion to the shape of the leaves.)—Our species are mostly annuals, flowering through late summer and autumn.

Chenopodium album L. Erect, *more or less mealy; leaves varying from rhombic-ovate to lanceolate* or the uppermost even linear, acute, *all or only the lower more or less angulate-toothed;* clusters spiked-panicled, mostly dense; calyx (2–2.7 mm. broad) with strongly carinate lobes, nearly or quite covering the seed.—Introduced everywhere. (Naturalized from Europe.)

PIGWEED

[*Amaranthus retroflexus* L.]

ONE OF the most robust, devil-may-care weeds is the Pigweed. "Careless weed" is another of its common names, and it comes by this name honestly enough. In good rich soil it cares for nothing. Wind, hail, fair weather and foul are all the same to the pigweed. Nor does it care what plants are its competitors. It can usually shoulder out any plant within reach, and it has a considerable reach. The name pigweed, however, has no reference to the piggish nature of the plant. It refers to the gustatory pleasure the weed affords pigs. Hogs will leave their corn to feast on pigweeds. In spite of its bristling appearance (it always reminds one of a boisterous young sailor with a week's growth of sandy beard on his face and his cap on the side of his head) the leaves are tender, and if the smacking of lips and satisfied grunts mean the same thing to pigs that they do to man the weeds must be delicious.

The pigweed is said to be "adventive from tropical America." Here again is shown its careless way. It may have originated in the tropics, but it is just as much at home in Canada as in Brazil. It is found all over the United States and Canada, and it is called in different places Amaranth pigweed, Green amaranth, Redroot, Rough pigweed, Chinaman's greens, Careless weed, and Pigweed.

The weed is a first cousin to the Thorny amaranthus and the

tumbleweeds; it is also related to the cockscomb, prince's-feather, and Joseph's coat. Sometimes one may see a pigweed that might well be used as a decorative plant because of its colors. The whole amaranthus family seems to be filled with the ability and the desire to vary, and the pigweed is no exception. The reason it can fit diverse environments so easily is because among the thousands of seeds each plant produces there is variation enough to permit some of the plants coming from those seeds to grow where the mother plant could never have grown. Some of the same seeds will produce plants with colored leaves; some will be wholly devoid of beauty, while others will be just as lacking in adaptability; but that is the way with variation. It is the spice of life, and the pigweed is chockful of it.

An explanation of the generic part of the name, *Amaranthus,* is given in the description of the Thorny pigweed. It means unfading. The specific part, *retroflexus,* means turning down or toward the stem, and refers to the drooping leaves of the pigweed which do turn down in a peculiar way. So the pigweed is the unfading plant with retroflex leaves.

As common as it is, the pigweed is not a bad weed. It is easily mastered. It is an annual, and although there may be as many as 115,600 seeds matured by a single plant, according to Doctor Beal, most of these will germinate in cultivated fields and gardens and go to feed the soil bacteria, since they are pulled into the soil by the plows and hoes that destroy the seedlings by the thousands. The few that survive may become fully grown plants, only to be chopped off with a hoe and carried to the pigsty. Such is a fitting end to the pigweed.

TECHNICAL DESCRIPTION

Amaranthus (Tourn.) L., see Thorny pigweed.
Amaranthus retroflexus L. Roughish and more or less pubescent; leaves dull green, long-petioled, ovate or rhombic-ovate, undulate; the *thick spikes* crowded in stiff *glomerate panicle; bracts awn-pointed,* rigid, exceeding the acute or obtuse sepals.—Cultivated grounds, common; indigenous southwestward. (Adventive from Tropical America.)

FIG. 28. The Pig Weed that is the Careless Weed

THORNY PIGWEED

[Amaranthus spinosus L.]

IF THE gardener does not know his pigweeds he is likely to rue his ignorance when he inadvertently attempts to pull up the thorny species. It looks very much like the true pigweed which has no thorns and pulls up readily enough. As a general thing one such mistake is all that is required to teach the weed eradicator to use more perspicacity when he delves among the pigweeds. There are two sharp, stiff spines in the axil of every leaf of this hateful weed, and it *is* hateful, for those spines are usually quite invisible when viewed from above—the only view for him who stoops to pull weeds.

The plant is a cousin of all the Amaranthus pigweeds, tumbleweeds and careless weeds, and of such beautiful decorative plants as the cockscombs, the prince's-feather, and Joseph's coat. These are all in the Amaranthus family, and all show a remarkable range of variation. Because of this innate ability of the Amaranthus to vary, some of the most spectacular decorative plants have been obtained by careful selection. But the only variation the thorny amaranth (thorny pigweed) seems to have is toward the satanic. Its horns get longer and sharper every time one's skin comes in contact with them.

This weed, like its cousin the true pigweed, is a lover of gardens and cultivated fields. It likes chicken lots and horse lots, too, and often uses these places to produce its millions of seeds for the near-by gardens and truck patches. It is more of a hot-weather plant than its cousin and so is more of a pest in the southern than in the northern States.

The botanical name of the plant, *Amaranthus spinosus,* is more descriptive than any of its common names. The fact is it has few common names, since it has been in most English-speaking communities a comparatively short time. It is from tropical America, and there are men still living, even in the South, who remember

Fig. 29. The Thorny Pigweed. Note the
thorns at the base of the leaves

its first appearance in their community. The botanical name *Amaranthus* is from the Greek and means the plant that never fades. It refers to the unchanged appearance of such weeds as the pigweed and to such decorative plants as the cockscomb. The *spinosus* part of the name scarcely needs interpretation. It is from the Latin and means filled with spines. "The spiny weed that does not fade," is the free translation of the name, and "Spiny amaranthus" should be the name in common use.

TECHNICAL DESCRIPTION

Amaranthus (Tourn.) L. Flowers 3-bracted. Calyx glabrous. Stamens 5, rarely 2 or 3, separate; anthers 2-celled. Stigmas 2 or 3. Fruit an ovoid 1-seeded utricle, 2-3-beaked at the apex, mostly longer than the calyx, opening transversely or sometimes bursting irregularly. Embryo coiled into a ring around the albumen.—Coarse annual weeds, with alternate and entire petioled setosely tipped leaves, and small green or purplish flowers in axillary or terminal spiked clusters; in late summer and autumn. (The Greek word meaning *unfading,* because the dry calyx and bracts do not wither.)

Amaranthus spinosus L. Smooth, bushy-branched; stem reddish; leaves rhombic-ovate or ovate-lanceolate, dull green, a pair of *spines in their axils;* upper clusters sterile, forming long and slender spikes; the fertile globular and mostly in the axils; flowers yellowish-green, small.— Waste grounds, Maine to Minnesota, and southward. (Naturalized from Tropical America.)

POKEWEED

[*Phytolacca decandra* L.]

THE POKEWEED is so different from any other weed that the most uninterested observers among rural people have a name for it. "Pokeweed" is probably the most used of all of its names, but it is also called Poke, Pokeberry, Virginia poke, Inkberry, Red-ink plant, Scoke (also spelled Skoke), Pigeon berry, Garget, Coakum, American cancer, Cancer jalap, Pocan or Pocan bush, American nightshade, and Chongras. The striking difference we see between these common names and those of other plants is that whatever the name may be it always applies to the Pokeweed. Other common names of plants are likely to refer to other species. "White-top," for instance, is used for several widely different weeds, but "Inkberry" means Pokeweed, and so do all the other names here given.

The size and shape of the plant are so different from the size and shape of other herbs that it would be strange if every one seeing it for the first time did not ask, "What is it?" The shape is like that of a small tree. Its rather long branches, supported by its hollow, trunklike stem which may be as much as four inches in diameter and eight feet in height, are characters that make a big pokeweed a very conspicuous plant. Since the root is perennial it must remain undisturbed for several years if the plant is to attain its maximum size. One usually finds the big pokeweeds towering above other weeds in some neglected fence row. Here their branches, bearing long clusters (racemes) of flowers and dark-red fruits, are spread out above and over the fence like the limbs of a small tree. Of course the branches are not woody, and the leaves are much larger and smoother than the leaves of the great majority of woody plants. And the pokeweed leaves are often mottled. The weed is very susceptible to a mosaic disease, and the presence of the disease shows up in the mottling of the leaves.

Pokeweed is listed among both the medicinal and the poisonous plants. Its roots and seeds contain poison enough to be deadly, but the young leaves are often cooked as greens. It is said that no dish of greens is quite complete that does not contain poke. The leaves are probably as poisonous as the roots and seeds until they are boiled, when the poisonous principle is steeped out and is poured off with the water when the greens are made ready for the table.

It is from the dried roots and fruit that the phytolacca sold at drugstores is made. This extract or tincture of phytolacca is said to be one of the best remedies known for reducing caking and swelling of the udders of cows. Farmers and dairymen make up

Fig. 30. The Pokeweed is one of the biggest of them

a mixture of phytolacca and lanolin (wool oil) and rub this into the ailing udders.

Pokeweed roots and berries are dried and sold on the market. Children make red ink out of the fresh berries. It is the red color that gives the *lac* to *phytolacca*. *Lac* is the same as the French word *laque* and the English word *lake*, both of which mean purplish-red. The *phyto* part of the word is the Greek name for plant, so *Phytolacca* is the purplish-red plant. There is a disagreement as to the specific name. Gray's *Manual of Botany* gives it as *decandra* L., while nearly all other authorities give it as *americana* L. Of course the *americana* means the American *Phytolacca*, while *decandra* means the ten-stamened *Phytolacca*. *Decandra* is from two Greek words, one meaning ten and the other meaning male organ. Since the male organs of a flower are stamens, *decandra* means ten-stamened.

TECHNICAL DESCRIPTION

Phytolacca (Tourn.) L. Calyx of 5 rounded and petal-like sepals. Stamens 5–30. Ovary of 5–12 carpels united in a ring, with as many short separate styles, in fruit forming a depressed-globose 5–12–celled berry, with a single vertical seed in each cell. Embryo curved in a ring around the albumen.—Tall and stout perennial herbs, with large petioled leaves, and terminal racemes which become lateral and opposite the leaves. . . .

Phytolacca decandra L. A smooth plant, with a rather unpleasant odor, and a very large poisonous root (often 1–1.5 dm. in diameter) sending up stout stalks at length 2–3 m. high; calyx white; stamens and styles 10; ovary green; berries in long racemes, dark-purple, ripe in autumn.—Low grounds and rich soil, southern Maine to Ontario, Minnesota and southward. July–September.

CARPET WEED
[*Mollugo verticillata* L.]

THE CARPET WEED is a sprawling example of designed humility. It is so humble that it is a genuine Uriah Heep among plants. It is one of those self-effacing, unobtrusive individuals that gets all and does all it was intended to get and do, and yet no one is aware that it is getting and doing it. The weed is seldom noticed by any one but the man with the hoe, and he sees it only as another weed, or if its persistence and shape awaken a thought as he hoes it into the soil, it is, "How can such a frail little thing become such a nuisance?"

Like the dandelion and other low-growing plants, the Carpet weed is constructed so that it can make and use the maximum amount of food with the expenditure of a minimum amount of energy. All of its leaves lie in the highest concentration of carbon dioxide to be found in the free atmosphere. The carbon from the carbon dioxide and the water from the soil are made into the sugar and starch that furnish the energy required by the plant in its growth and seed production. How rapidly this energizing food can be made and used only he who knows some of the sprawling weeds can appreciate. Those who have fought dandelions know something about it, but the dandelion is scarcely comparable, for it stores its food in its reservoir of a root from which it sends up a shoot that blooms and seeds in a very few hours. The Carpet weed does not go to so much trouble. It simply lies there making food and seed, and then dies, depending upon the wind that bloweth where it listeth to carry far and wide its exhausted branches filled with seeds.

Carpet weed is the most descriptive name for the plant, but it has been called Indian chickweed, Whorled chickweed, and Devil's grip. The generic name, *Mollugo,* is derived from the Latin and the French *mollis* meaning soft. *Verticillata,* the specific name, is derived from the Latin and means whorled. It refers more to the shape of the entire plant than to its whorled leaves, for the plant itself is a whorl.

The weed is said to be a native both of tropical America and of Africa. It is now found locally over most of North America east of the Rockies. A description of it is scarcely necessary. Suffice it to say that the lacy, sprawling plant with its leaves in whorls at the swollen joints of its stems is the Carpet weed. It might be confused with pursley or with the spotted spurge, both of which

FIG. 31. The Carpet Weed as seen from above

have much the same habit of growth as the Carpet weed, but pursley has smooth, thick stems and leaves, and the spurge has a milky juice. All three of these plants are likely to be found together, as they are all three lovers of the rich soil of cultivated fields and gardens. The Carpet weed lies closer to the ground than the other two, its stems are swollen at the nodes, and its leaves are in whorls. It is, therefore, easily distinguished from the others by him who will bear these facts in mind.

TECHNICAL DESCRIPTION

Mollugo L. Sepals 5, white inside. Stamens hypogynous, 5 and alternate with the sepals, or 3 and alternate with the 3 cells of the ovary. Stigmas 3. Capsule 3-celled, 3-valved, loculicidal, the partitions breaking away from the many-seeded axis.—Low homely annuals, much branched; the stipules obsolete. (An old Latin name for some soft plant.)

Mollugo verticillata L. Prostrate, forming mats; leaves spatulate, clustered in whorls at the joints, where the 1-flowered pedicels form a sort of sessile umbel; stamens usually 3.—Sandy river-banks, roadsides, and cultivated grounds. June–September. (Immigrant from farther south.)

CHICKWEED

[*Stellaria media* (L.) Cyrill.]

ONE OF the meanest of the lawn weeds is that little, crawling member of the pink family known as the chickweed. Filled with a vitality even superior to that of bluegrass, it is on the job as soon as the frost is out of the ground, and long before the grass is ready to cut this pest is in full bloom and spreading itself.

It is too bad there is not a common name for the plant that will express the disgust of the lawn maker when he finds it has moved in on him. It ought to be called "lawn whoosh" or something like that, but it is not. It has always been seen when named through the eyes of nature lovers, so we have as common names

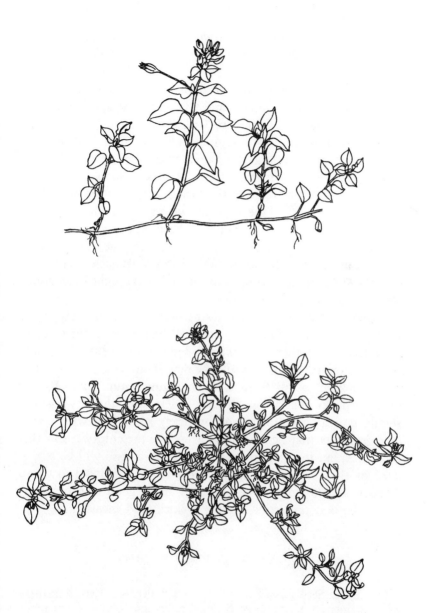

FIG. 32. Two views of Chickweed, a pest of the lawn

Starweed, Star chickweed, Starwort, Bind weed (a name remotely descriptive), Winter weed, Satin flower, Tongue grass,
and Mouse ear. Even the botanical name is heavenly. *Stellaria
media* (L.) Cyrill. means the medium, star-flowered weed whose
genus was named by Linnæus and whose specific appellation was
changed to *media* by Domenico Cirillo.

Chickweed is said to be found all over the world. It is listed as
a medicinal plant, but its medicinal virtues are few and insignificant in comparison with those of other plants. The name
Chickweed seems to refer to the fact that wild birds as well as
domestic fowls feed on the seeds and plants. Hogs and rabbits are
fond of the weed, but it is said that sheep do not care for it and
that goats refuse to eat it. It is also said that the young leaves
boiled can scarcely be distinguished from spinach. Gathering
chickweed leaves for boiling must be one of the light occupations
we used to hear so much about.

In many places chickweed has become one of the very worst
lawn weeds. Often liming the soil will help keep it out of bluegrass lawns. Another good way to fight it is to dust the plants
with ammonium sulphate while the dew is on. This will burn the
plants enough to kill them, and since ammonium sulphate is a
good fertilizer for lawns, all that is left after the day's action
should be washed into the soil by heavy sprinkling with water.
The grass tops may be burned off as well as the chickweed by this
treatment, but the grass roots are perennial and so will be stimulated by the ammonium sulphate that is washed down to them.
A good sod should thus be produced and a good vigorous grass
sod is the best protection against a chickweed invasion.

TECHNICAL DESCRIPTION

Stellaria L. Sepals 4–5. Petals (white) 4–5, deeply 2–cleft, sometimes
none. Stamens 8, 10, or fewer. Styles 3, rarely 4 or 5, opposite as many
sepals. Pod ovoid, 1–celled, opening by twice as many valves as there
are styles, several many-seeded. Seeds naked.—Flowers solitary or
cymose, terminal or appearing lateral by the prolongation of the stem

from the upper axils. (Name from *stella,* a star, in allusion to the star-shaped flowers.)

Stellaria media (L.) Cyrill. *Annual or nearly so; stem hairy in lines; leaves ovate* to ovate-oblong, the lower on hairy petioles; petals shorter than the calyx, 2-parted; stamens 3-7; seeds scarcely roughened.—A common weed. (Naturalized from Europe.)

CORN COCKLE

[*Agrostemma githago* L.]

CORN COCKLE is a pretty "pink flower that grows in the wheat." It is pretty enough, but it has a notorious reputation. Its seeds ripen simultaneously with winter wheat, and since they are about the same size and weight of wheat grains their removal from wheat seed is very difficult. The weed is a winter annual just as winter wheat is, and for this reason it is sure to be sown with the wheat if it is thrashed with it. Farmers who want to rid their fields of cockle should take the only sure way of doing it: examine every inch of the fields containing the pest before the wheat is ready to harvest, and remove every Corn cockle stalk found, roots and all. In most fields the plant is very easily pulled up.

The weed has several names but none of them is very descriptive, unless it be "Corn rose." It should be remembered that this, like so many of our weeds, brought its names with it from England, and that corn to the Englishman is wheat, not maize. So Corn rose, Corn cockle, Corn campion, and Corn mullein all refer to the wheat in which this weed grows. Crown-of-the-field and Old Maid's pink are two other names that appraise the weed's beauty rather than its worth.

For Corn cockle is not worth much. It finds a place in the herbals as a medicinal plant, but the very same seeds that make it medicinal vex the wheat grower and poison his stock. The medicine made from the seeds by trituration is said to be useful in treating paralysis and gastritis, two very dissimilar diseases, but the herbals declare such to be the case.

One seldom sees Corn cockle anywhere except in wheat fields, and for this reason it is of little interest to any one but the botanists and the growers of winter wheat. Its classification interests the botanist. The plant is a true pink. It belongs to the family *Caryophyllaceæ,* the same family that contains the carnation and all of its beautiful relatives, but it is not *pinked* like the carnation. It is not the color of the flower that gives to the pinks their name but the pinking of the edges of the petals. The color, pink, came into use just as the colors rose and violet came. The color is the color that is usually found in pinks, and pinks are the flowers that are usually pinked.

The common name, pink, then, is easy of derivation, but not so the botanical word, *Caryophyllaceæ,* which stands as the name of the pink family. It requires the Oxford Dictionary and some mental gymnastics to get at the meaning of that word. Seemingly Reichenbach, who gave the family its name, turned several mental somersaults and landed on *Caryophyllaceæ* when he smelled the Gillyflower (pronounced jilly-flower). But he really did not. The Gillyflower smells like cloves and is called the clove pink. Its botanical name is *Dianthus caryophyllus. Caryophyllus* comes from the Greek word *Karyophyllon,* the name of the clove tree. So when a good name from a dead language was needed to designate a family of plants containing several spicy-smelling flowers, and especially one that smelled like cloves, *Caryophyllaceæ* was the natural conclusion.

The generic name, *Agrostemma,* is from two Greek words meaning field and crown; the specific name, *githago,* refers to another plant, *gith* or *git* (*Negrella sativa*), whose seeds are also used in medicine.

Thus it is seen that Corn cockle, the "little pink flower that grows in the wheat," interests the botanist as well as the wheat grower but for reasons widely different.

TECHNICAL DESCRIPTION

Agrostemma L. Calyx ovoid, with 10 strong ribs; the elongated teeth (in ours 2–3 cm. long) exceeding the 5 large unappendaged petals. Stamens 10. Capsule 1–celled. Leaves linear.—Tall silky annual

Fig. 33. Corn Cockle, a pink, but a bad one

or biennial. (Name from two Greek words meaning field and crown.)

Note: According to the newer dictionaries the word is a combination of the new-Latin word *agro* meaning field, and the Greek stem meaning garland.

Agrostemma githago L. Flowers 2.5–4 cm. in diameter; petals purplish-red, paler toward the claw and spotted with black.—Grainfields, and less frequently by roadsides. (Introduced from Europe.) Seeds poisonous.

BOUNCING BET

[*Saponaria officinalis* L.]

THERE is no plant more suggestive of old colonial days than the Bouncing Bet. The stout little pink with its modest color and its profusion of rather persistent flowers must have come near pleasing the puritanical and thrifty natures of those early New England housewives.

Bouncing Bet might well be called a beautiful weed. It can be, and is being used, as a decorative plant, but it has some very weedy proclivities. It has the tenacity of the weediest of weeds; it also has considerable aggressiveness when unmolested by man. It is often found haunting a home site so long after its admiring planter left the scene that no memory of him or even of his dwelling remains, and yet, in her *Nature's Garden* Neltje Blanchan can truthfully say, "A stout, buxom, exuberantly healthy lassie among flowers is Bouncing Bet, who long ago escaped from gardens whither she was brought from Europe, and ran wild beyond colonial farms to roadsides, along which she has travelled over nearly our entire area." She certainly stays put and as certainly goes.

Bouncing Bet is only one of the plant's names. Being English, or at least having lived long in England, it might be expected that a plant so popular would be much christened. "Bouncing Bet" seems to be the best known of any of the names among which are Soapwort, Sweet Betty, Wild Sweet William, Scourwort, Old Maid's pink, London pride, Hedge pink, Boston pink, Chimney pink, Sheepweed, Soapwort gentian, World's wonder, Lady-at-the-gate, Wood phlox, and Mock-gilliflower.

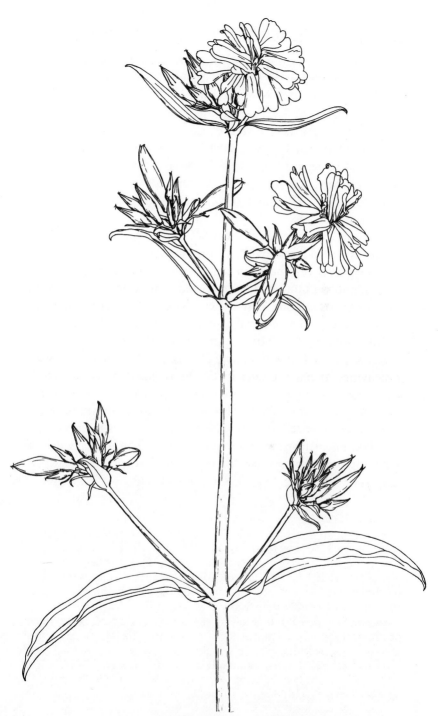

Fig. 34. A double-flowered Bouncing Bet

The names referring to soap are the most descriptive. The bruised leaves and stems churned about in water will make suds. The juice of the plant is somewhat poisonous, and it is said that "a decoction of the leaves will cure itch." It was the widespread belief in the medicinal virtues of Bouncing Bet that placed it in so many of the early home yards. It was used to treat venereal diseases in those good old days. Modern herbals say it should be used with care, intimating that the remedy is sometimes worse than the disease. The saponin content of the weed is very poisonous to some people.

Any one can know Bouncing Bet even before it blooms if he has but observed it once while it was in bloom. The heavy, smooth, opposite leaves that have no stems (petioles) but clasp the stem of the plant and unite with each other around it in a rather unique way make recognition easy. Then there is the soap test. If the broken leaves and stems make suds or something like suds when shaken in water the plant is surely Bouncing Bet.

The genus to which the plant belongs gets its name, *Saponaria,* because of the soapy nature of this particular species. Its specific name, *officinalis,* refers to the apothecary's shop. The words comes from the Latin through the French with about the same origin that the word office has. *Officinalis* means that this plant was prepared and sold from an apothecary's shop. The name is used as a specific name for several different medicinal plants.

TECHNICAL DESCRIPTION

Saponaria L. Calyx narrowly ovoid or subcylindric, 5-toothed, obscurely nerved, naked. Stamens 10. Styles 2. Pod 1-celled, or incompletely 2–4-celled at base, 4-toothed at the apex.—Coarse annuals or perennials, with large flowers. (Name from *sapo,* soap, the mucilaginous juice forming a lather with water.)

Saponaria officinalis L. Flowers in corymbed clusters; *calyx terete;* petals crowned with an appendage at the top of the claw; leaves oval-lanceolate.—Roadsides, etc. July–September.—A stout perennial, with large rose-colored flowers, commonly double. (Adventive from Europe.)

PURSLEY
[*Portulaca oleracea* L.]

PURSLEY is the smooth, thick-stemmed, sprawling weed that is found in all rich gardens and cultivated fields throughout the United States and Canada after the summer has fully arrived. Strange as it may seem, this plant was once grown as a salad plant, and was also used by the ancients as a pot herb. Some cooks having a taste for late greens still use it mixed with turnip tops and such mustards as happen to be in season.

The fact that other plants have displaced Pursley as a food argues only that the other plants, lettuce, spinach, kale, etc., are better than it is in flavor perhaps, perhaps in nutrition, but not in vigor nor as a food for swine. The pigweed itself elicits no more grunts and lip smackings from the pigs than does a bunch of Pursley. It might well be called "pig salad" or "pig relish," for its common names are few, and all it has seem to be corruptions of its generic name, *Portulaca*. How Pursley, Purslane, and Pussley were derived from Portulaca is rather hard to see. Purslane was probably the first corruption, and it came by the way of the French. An attempt to spell what the Frenchman said when he pronounced the Latin word, Portulaca, brought the word Purslane. Then, of course, the English had to modify that, and so Pursley and Pussley are corrupted Purslane.

The weed is so much of a hot-weather plant that it never molests the early gardeners, but it can become quite a pest in the rich soil of gardens and cultivated fields after the middle of June. Like all low-growing plants it develops rapidly and seeds quickly. Its succulent, rapid-growing nature makes of it a good soil builder and an excellent hog feed. So he who has this weed to contend with should have as an eradication slogan, "Feed it to the hogs or to the soil bacteria."

The beautiful flowering moss is a portulaca, *Portulaca grandiflora,* and this weed is one of its first cousins. Portulaca is the ancient name for such thick-leafed, thick-stemmed plants with flowers made up as these plants' flowers are. The specific name of

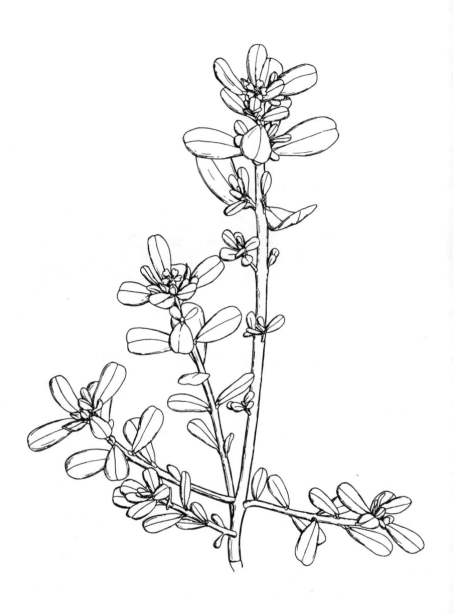

Fig. 35. The Pursley. It was
once used as a salad plant

the weed, *oleracea,* refers to the fact that the plant is used as a pot herb. And so *Portulaca oleracea* L. means the edible portulaca named by Linnæus.

TECHNICAL DESCRIPTION

Portulaca (Tourn.) L. Calyx 2-cleft; the tube cohering with the ovary below. Petals 5, rarely 6, inserted on the calyx with the 7-20 stamens, fugacious. Style mostly 3-8 parted. Pod 1-celled, globular, many-seeded, opening transversely, the upper part (with the upper part of the calyx) separating as a lid.—Fleshy annuals, with mostly scattered leaves. (An old Latin name, of unknown meaning.)

Portulaca oleracea L. *Prostrate,* very smooth; *leaves obovate* or wedge-form; flowers sessile (opening only on sunny mornings); sepals keeled; petals pale yellow; *stamens 7-12; style deeply 5-6-parted;* flower-bud flat and acute.—Cultivated and waste grounds; common.—Seemingly indigenous westward and southwestward. (Naturalized from Europe.)

SMALL-FLOWERED BUTTERCUP
[*Ranunculus abortivus* L.]

THE WORD buttercup does not signify "weed" to the minds of most people. "The buttercup, the little children's dower" is a relative of the subject of this sketch, but only he who knows plants as the botanist knows them would be likely to discover the relationship. A buttercup to the layman is a plant whose flower petals are yellow and glossy, as if smeared with that oleaginous substance. To the botanist it is a plant belonging to the genus *Ranunculus,* the crowfoot genus, the genus that gets its scientific name from the place it grows: among the little frogs, and its English, Crowfoot, from the fact that the most of the species of the genus have leaves that resemble to a greater or less degree a crow's foot.

This small-flowered buttercup belongs in the buttercup (crow-foot) genus even though its flowers are not buttery and its lower leaves are more like fans than crows' feet. *Ranunculus abortivus,* its botanical name, is its best name. It means the plant that grows where little frogs are and whose flowers are abortive, that is, whose flowers lack petals or come very close to lacking them. It has one other English name which fairly well describes the plant when one knows it to be a crowfoot. It is the "Smooth-leaved crowfoot." The weed is rather common in shady and moist places. Like most of the buttercups it appears early. It is one of the spring weeds, and in some places it becomes a genuine nuisance, especially in the wet meadows and pasture fields of the eastern part of the United States. It seeds quickly and abundantly, and infested fields and lawns should be plowed if possible before the plant blooms, or should be closely mowed as soon as the first flowers appear. These are the two most effective ways of eradicating the weed.

To return to the plant's relatives: the family of the small-flowered buttercup is a large one, and although this plant is the only member of the family treated in this book the reader should know that the larkspurs, the anemones, hepaticas, the columbine, and clematis, as well as all of the buttercups, belong in this family.

TECHNICAL DESCRIPTION

Ranunculus (Tourn.) L. Annuals or perennials; stem-leaves alternate. Flowers solitary or somewhat corymbed, yellow, rarely white. (Sepals and petals rarely only 3, the latter often more than 5. Stamens occasionally few.)—(A Latin name for a little frog; applied by Pliny to these plants, the aquatic species growing where frogs abound.)

Ranunculus abortivus L. Biennial, slightly succulent; stem 1.5–6 dm. high, covered with a short sparse sometimes fugacious pubescence; primary *root-leaves round-heart-shaped with a wide shallow sinus or kidney-form,* barely crenate, the succeeding often 3–lobed or 3–parted; those of the stem and branches 3–5–parted or divided, subsessile, the divisions oblong or narrowly wedge-form, mostly toothed; *petals pale yellow, shorter than the small reflexed calyx; receptacle villous;* carpels minute, merely mucromulate.—Shady hillsides and along brooks, common. April–June.

FIG. 36. The Small-Flowered Buttercup

PEPPERGRASS

[*Lepidium virginicum* L.]

THIS little weed goes to market much too often to please those interested in lawn seed. As a weed it is of more importance in lawns than anywhere else, and much too frequently it makes its entrance by way of the seed sower. It does not have to get in that way, however. There are nearly always a few stray, seed-filled stalks of Peppergrass close at hand, and if there are not, the tiny seeds are carried great distances by gusts of wind and by whirlwinds. Many times, too, the entire plant breaks off and plays tumbleweed, scattering seeds all along the way, and when such a tumbler crosses a lawn every bare spot it passes over will break out with peppergrass rosettes.

Like Shepherd's purse, with which its rosettes are often confused, it starts its semi-biennial existence in the fall. If these beginning rosettes are not disturbed the stems with their many branches will arise and mature their many seeds by early May. It is these wiry stems with their wiry branches that make it the ugly lawn weed it is, as well as the joy of the seed-eating birds.

The best way to rid a lawn of the pest is to spud out the rosettes in late fall or early spring. The rosettes of the Shepherd's purse are likely to be there too, and since both of these weeds are mustards and give zest to a mess of greens, why not chuck them in a pot along with some dandelion and dock leaves that are sure to be spreading themselves near by, if not in the same lawn? There is nothing so innately satisfying as to be able to eat one's enemy. A dish of greens is savory when eaten with the knowledge that it has removed a lawn blemish.

Peppergrass has several other common names: Bird seed, Bird's pepper, Hen pepper, Tongue-grass, and Poor-man's pepper. The seeds are relished by birds and so it deserves the names Bird seed and Bird's pepper. Its botanical name means the plant with the little scales (fruits) that grows in Virginia, but it grows everywhere now. It is a weed with a virtue or two, but still a weed.

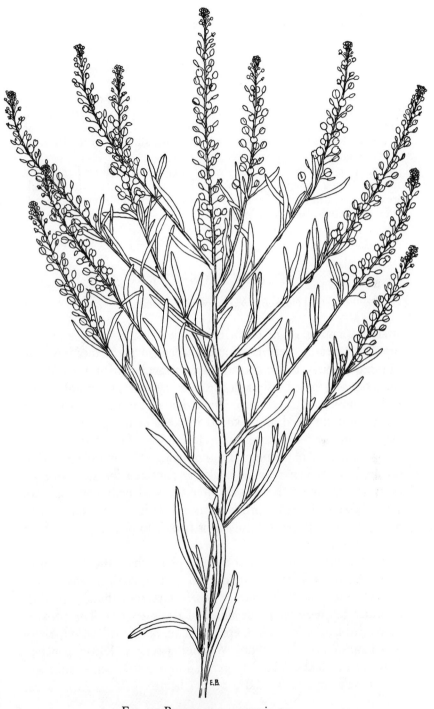

FIG. 37. Peppergrass, a cousin to
the Shepherd's Purse

TECHNICAL DESCRIPTION

Lepidium (Tourn.) L. Pod roundish, much flattened contrary to the narrow partition; valves boat-shaped. Seeds solitary in each cell, pendulous. Cotyledons incumbent, or accumbent. Flowers small, white or greenish. (Name from the Greek, *a little scale,* alluding to the fruit.)— Ours are annuals or biennials, except the last.

Lepidium virginicum L. Cotyledons *accumbent* and seed minutely margined; *pod marginless* or obscurely margined at the top; petals present, except in some of the later flowers.—A common weed of roadsides and waste places. June–September.

SHEPHERD'S PURSE

[*Capsella bursa-pastoris* Medic.]

SHEPHERD'S PURSE is one of the commonest of the spring weeds. It appears in nearly every lawn and garden, in every meadow and pasture, in every truck patch and cultivated field, not as a bad weed but as a sort of visitor. It comes everywhere, seemingly filled with curiosity, and its fate is often that of the curious cat. Into the ground are hoed and plowed myriads of the plant's rosettes that began their existence either late in the fall or during the first warm days of spring. The thousands of mustard seeds that fill the "purses" on a single stalk are scattered far and wide by every gust of wind that passes after the seed pods have opened. This accounts for the spring invasion. This is why Shepherd's purse, in company with a cousin or two, is to be seen everywhere in the spring of the year.

There never was a better-named plant than the Shepherd's purse. Even the botanical name, freely translated, means "Shepherd's purse." *Capsella bursa-pastoris* says something like this: the little capsule that is the purse of the shepherd. The plant is also called by the English Shepherd's bag, Shepherd's script, Shepherd's sprout, Lady's purse, Witches' pouches, Rattle pouches, Case weed, Pickpocket, Pick-purse, Blind weed, Pepper-and-salt, Poor-man's parmacety, and Mother's heart.

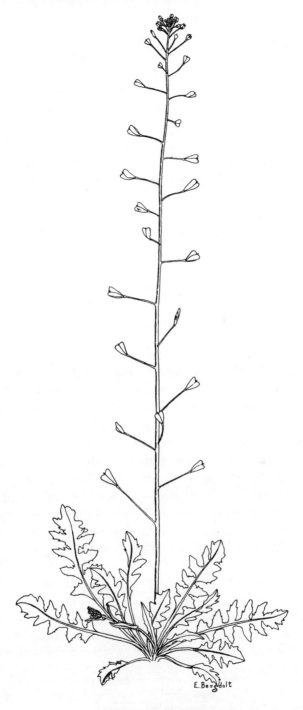

FIG. 38. The Shepherd's Purse with
a whole line of purses

Shepherd's purse belongs to the mustard family, and like many of the mustards it has salad properties. It is often used in "greens" along with other mustards, dandelions, wild lettuce, and poke. It is classed as a medicinal plant and "when dried and infused, it yields a tea that is still considered by herbalists one of the best specifics for stopping hemorrhages of all kinds—of the stomach, the lungs or the uterus and more especially bleeding from the kidneys."

Every one should know the Shepherd's purse, not so much because of its merits or demerits, but because, like Mother Goose rhymes, it belongs to child life and to the folkside of our nature. Fairyland is close to every stalk strung with its triangular purses. Every one should know the weed, but he must not expect too much or he will never find it. It is a humble little plant. Fully grown it is never more than eighteen inches high and the average stems are only about half that high. It owes most of its reputation to those purselike seed pods; they are so different from any other seed pods. They are "obcordate-triangular," the technical description says, and the "obcordate" means heart-shaped backwards. That is, the point of the heart is toward the stem instead of away from it, as it is in the case of cordate leaves.

Find a plant that has a wirelike green stem with few leaves on it but with a great number of little heart-shaped seed pods in a loose spike, and at its upper end you will see the Shepherd's purse in its most easily identified stage. It is not so easily identified as a rosette of basal leaves, or even before the seed pods begin to form on the spike. Learning the life history of Shepherd's purse is like digging a well: one has to begin at the top.

TECHNICAL DESCRIPTION

Capsella Medic. Pod obcordate-triangular, flattened contrary to the narrow partition; the valves boat-shaped, wingless. Seeds numerous. Cotyledons incumbent.—Annual; petals small, white. (Name a diminutive of *capsa,* a box.)

Capsella bursa-pastoris (L.) Medic. Stem-leaves arrow-shaped, sessile. Common weed; April–September. (Naturalized from Europe.) Extremely variable in foliage and outline of pod.

WILD MUSTARD

[Brassica arvensis Ktze.]

ONE OF the meanest weeds of small grain fields is what is known as Wild mustard, Field mustard, Charlock, Field Kale, Kedlock, Kerlock or Curlock, Bastard-rocket, Runch weed, Crowd weed or Kraut weed, and Yellow flower. These are the names of the weed in different localities, and they are given so that he who reads this may know what is meant when an enraged grain grower curses a plant bearing any one of these names.

In many sections of the country small grain farmers do not need a book to tell them what the weed is; but there are sections still free from it and still others where it has but recently arrived. One purpose of this sketch is to warn those who do not know that the little yellow-flowered plant which appears in a wheat, oat, barley or rye field, or even in an alfalfa field, is likely to be the hated Wild mustard.

Where it has become established the plant is certainly a pernicious weed. How pernicious it is may be gathered from the fact that it costs the California farmer from two to four dollars an acre to control it in his fields of small grain. The best means of destroying the weed found thus far seems to be the 10 per cent (by weight) solution of sulphuric acid used as a spray. It requires special equipment and much work to put on this spray, but when the field becomes badly infested it is either spray or lose the crop. The weed grows so much faster than wheat, oats, or barley that the grain plants are soon smothered out by its heavy foliage.

Since there are localities in the United States where the weed has never appeared and others where it has but recently come on the scene, there must needs be some sort of description of the pest given here. To farmers and gardeners of those favored localities let it be said, "Beware of mustard plants with yellow flowers and numerous siliques." What are siliques? They are the long

FIG. 39. Wild Mustard, a very bad weed
in small grain fields

seed pods of some of the mustard plants; for instance, the seed pods of the garden radish are siliques. Let it then be said to those who have never seen this particular weed: "Know it is time to begin to fight when you find in your wheat, oat or barley field an early developing, much-branched, yellow-flowered, silique-filled weed. If you make such a discovery and do not want to be harassed by it the rest of your life, you will remove the culprit at once, and not only remove it but destroy it with fire. If you will pull up, carry out, and burn every one of the plants that appears in your fields you will be relieved in the future of one of the worries visited upon those who saw the advent of the plant but regarded it as just another weed."

The best name the plant has is "Field mustard." Its botanical name means Field cabbage. *Brassica* was the ancient name for cabbage, and *arvensis* refers to cultivated fields. The great number of common names applied to it shows it to be an imported as well as an important weed.

TECHNICAL DESCRIPTION

Brassica (Tourn.) L. Pod slender or thickish, nearly terete or 4-sided, with a stout often 1-seeded beak; valves 1–5-nerved. Seeds globose, 1-rowed. Cotyledons conduplicate.—Annuals or biennials, with yellow flowers. Lower leaves mostly lyrate, incised, or pinnatifid. (The Latin name of the Cabbage.)

Brassica arvensis (L.) Ktze. *Knotty pods fully one third occupied by a stout 2-edged beak;* upper leaves rhombic, scarcely petioled, merely toothed; *fruiting pedicels short, thick; pods smooth or rarely bristly,* 4 cm. long.—Noxious weed in grainfields, etc. (Naturalized from Europe.)

WHITE CLOVER

[*Trifolium repens* L.]

WHITE CLOVER is usually regarded as anything but a weed. Its virtues rather than its faults impress nearly every one who has anything to do with the plant. Its fragrance and the simple beauty of its flowers linger in the memory. Braided into chains to form sweet-smelling garlands and leis for sunny childhood days, it blooms on memory's hills and dales long after the slobbering horses that fed on those hills and dales have been forgotten.

If a weed is a plant out of place, then almost any plant can get that way, and white clover is no exception. It is when white clover gets into a rock garden, a perennial flower bed, or a strawberry patch, or when all the horses on the place come in from the pasture drooling like babies, that the gardener and farmer call white clover a weed, and sometimes with emphasis.

The plant grows all over the United States and in nearly every country of the world. It is one clover that is not very particular about soil conditions. Whether it is sown or not, nearly every lawn and grassy strip along the streets and highways of the country can show a goodly sprinkling of white clover. One of the plant's virtues is its ability to fill in the spaces of a sparse bluegrass sod. According to the United States Department of Agriculture, from two to three million pounds of white clover seed are consumed by the United States every year, and the greater part of that amount is sown on lawns, and yet it may become a lawn weed. Every space left open, if the crab grass and chickweed do not get there first, will be seized by the white clover, or by one of its near relatives, either the black medic or the yellow hop clover.

Long before the flowers bloom white clover may be identified by its little, almost round leaflets which fold along the conspicuous midribs at night, and by the creeping stems which give to the plant its specific name, *repens*. The leaves, like all clover leaves, are compound, there being three leaflets to each leaf, and the manuals of botany call the shape of these leaflets *obcordate*, meaning that the notch of the heart is at the outer end of the leaflet. As a general thing the young leaflets are very nearly round, and these with the creeping stems that send out roots at every node make identification certain. When in bloom the Alsike clover looks very much like white clover, but its stems do not creep and so stand up higher than do the white clover stems. Its blooms are a little larger and have a little more color than is usually seen in

FIG. 40. White Clover with ripe head at the right
and a ripening head above it

white clover, but white clover blooms may have a pink tinge also. When one knows that the clovers have been called clover for hundreds of years, that in Anglo-Saxon the name was *clafre* and that in Middle English it was both *claver* and *clover,* it seems strange that this little plant should be called anything but white clover. Such is the case, however. It is called White Dutch clover, Honeysuckle clover, White trefoil, Purple grass (the older leaves have a purplish cast), Purplewort, Honey stalk, Lamb sucklings, and occasionally Shamrocks.

The botanical name, *Trifolium repens* L., is of Latin origin and means the creeping three-leafed plant. The generic name of the clovers is *Trifolium:* three leaves. Most of them were named by that famous namer of plants, Carolus Linnæus, and so L. follows nearly all of the botanical names of the clovers.

TECHNICAL DESCRIPTION

Trifolium (Tourn.) L. Calyx persistent, 5-cleft, the teeth usually bristle-form. Corolla mostly withering or persistent; the claws of all the petals, or of all except the oblong or ovate standard, more or less united below with the stamen-tube; keel short and obtuse. Tenth stamen more or less separate. Pods small and membranous, often included in the calyx, 1–6-seeded, indehiscent, or opening by one of the sutures.— Tufted or diffuse herbs. Leaves mostly palmately (sometimes pinnately) 3-foliate; leaflets usually toothed. Stipules united with the petiole. Flowers in heads or spikes. (Name from *tres,* three, and *folium,* a leaf.)

Trifolium repens L. Smooth perennial; the slender *stems spreading and creeping; leaflets inversely heart-shaped* or merely notched, obscurely toothed; stipules scale-like, narrow; petioles and especially the peduncles very long; the heads small and loose; *calyx much shorter than the white corolla;* pods about 4-seeded.—Fields and copses, everywhere; indigenous only in the northern part of our range, if at all. (From Eurasia.)

SWEET CLOVER

[*Melilotus alba* Desr.]

THE KING of the weeds, the best weed that grows, is the white sweet clover. It is the one weed that has been used as weeds should be used: to feed the soil bacteria. Of course it is of more use than ordinary weeds are for this purpose, since it contains nitrogen not taken from the soil. It contains no more nitrogen, however, than do any of the other succulent weeds of its size, but the nitrogen it contains is brand-new; taken from the air. For this reason the nitrogen supply in the soil is left untouched while the sweet clover is growing, and it is increased when the plant is turned under. Ordinary smartweeds, or any other weeds, will make available just as much nitrogen per unit weight of tops as will the sweet clover, but the smartweeds leave the ground with less nitrogen in it if they are not turned under, and they put no more back into the soil when they are turned under than was there before them, in some form or other, while the sweet clover will leave the soil a little better than it was, because of its roots, even if it is not used as a green manure.

How can it do this? It feeds some nitrogen-fixing bacteria that live in nodules (little knots) on its roots, and these bacteria give to the plant as fast as they die (and they die rapidly) all the nitrogen they have fixed. It is impossible to make living cells without nitrogen, and it is impossible for a plant, or any other living thing, to grow without the addition of cells. The size of the plant depends upon the number of its cells, and the number of its cells depends upon the nitrogen supply. Sweet clover is one of those legumes that has worked out the problem of getting its nitrogen indirectly from the air by feeding the dwarfs that know the secret of taking it from the air. Weed though it is (in some places it is a genuine nuisance) it is a wonderful soil builder for him who knows how to use it.

Sweet clover insists upon sweet soil, not so much because of itself as because it cannot live without its nodule-inhabiting bacteria, and they absolutely refuse to live in sour soil. This is why sweet clover does not become a bad weed in most fields. The soil is too sour there for its dwarf assistants. Where it does become a weed the soil is not in need of lime.

The plant has more virtues than that of soil building. It furnishes one of the best bee pastures and is often called "bee clover." The generic part of its botanical name, *Melilotus,* is made up of two Latin words. The *mel* part means honey, and the *lotus* part refers to a legume. Some of its common names go far enough back to have the Latin origin. They are Melilot, White melilot, and Corn melilot. Other common names are King's clover, Plaster clover (referring to the fact that it grew best where land plaster had been spread on the land), Sweet Lucerne, Wild laburnum, and Hart's clover. The last name refers to the fact that the hart (a deer) fed on the weed. This is another of its virtues: it makes good pasture, after cattle have learned to eat it, and it also makes good hay. It is interesting to know that in the sixteenth century hart's clover, a comparatively new weed in England at that time, was being indorsed by some of the land-holding gentry. They knew it to be a weed and some condemned it, but its advocates declared it to make both excellent pasture and hay.

Sweet clover is easily identified. It is the tall, bushy weed with the little white blossoms seen growing in almost brushlike thickets along so many highways. The entire plant is fragrant and can be identified by its odor alone. At a distance it is truly fragrant; close to the nostrils it is rather nauseating. Sweet clover and alfalfa look very much alike in their early stages, but one has only to smell of their crushed leaves to know which is which. The yellow sweet clover often seen with the white has the same odor as the white. Alfalfa is almost odorless in comparison.

Before giving the technical descriptions it might be well to translate the entire botanical name *Melilotus alba* Desr. In English it means the white honey lotus named by Desrousseaux.

FIG. 41. Sweet Clover, the prince of weeds

TECHNICAL DESCRIPTION

Melilotus (Tourn.) Hill. Flowers much as in *Trifolium,* but in spikelike racemes, small. Corolla deciduous, free from the stamen tube. Pod ovoid, coriaceous, wrinkled, longer than the calyx, scarcely dehiscent, 1–2-seeded. Annual or biennial herbs, fragrant in drying, with pinnately 3-foliate leaves. (Named from the Greek words *Mel,* meaning honey, and *Lotus,* meaning some leguminous plant.)

Melilotus alba Desr. Tall; leaflets narrowly obovate to oblong, serrate, truncate or emarginate; *corolla white,* 4–5 mm. long, *the standard longer than the other petals;* pods 3–4 mm. long, somewhat reticulate.—Rich soil, roadsides, etc., common. (Naturalized from Europe.)

WILD BEAN VINE

[*Strophostyles helvola* Britton]

THERE may be found growing in almost any poor neglected field from the east coast of the United States to the dust bowl, if not to the Rocky Mountains, a little wild bean vine that should have far more attention than it ever gets. It is one legume whose symbiotic bacteria do not shun acid soils. What are symbiotic bacteria? They are those bacteria that live in nodules on the roots of legumes, giving to the legume on which they grow the nitrogen the plant must have for its growth, but taking from the plant the sugar that they, the bacteria, must have for their own life processes. That is *symbiosis:* a living together for mutual benefit.

Well, the wild bean, unlike the most of the clovers and cultivated beans, has a strain of bacteria that can grow in very sour and very poor soil. Oftentimes the farmer discovers that a field he thought was worn out makes a surprising comeback after a rest of two or three years. He cannot understand it, but had he walked over that field with his eyes open before he plowed it he would have found it fairly well covered with either this species of wild bean vine or with a near relative, or with both species. Of course there were other weeds there, too, and if he plowed the field so as to get the most good out of those weeds the soil would have been benefited even if the bean vines were not there;

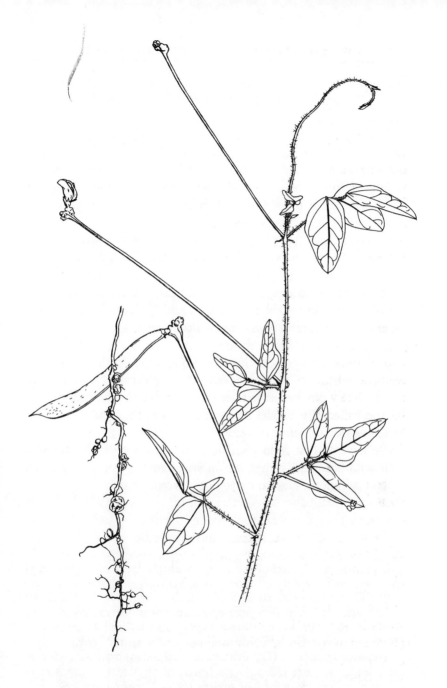

Fig. 42. The Wild Bean Vine and one of
its nodule-laden roots

but nine times in ten the surprising comebacks of such fields are due to the wild bean vines, and one look at the nodule-covered roots answers the question: why?

This is a weed that should be cultivated as a soil builder. It is an annual, but its little furry seeds should be planted in the late fall just as nature plants them. It might be possible to find a variety of this bean that would rival the sweet clover as a soil builder, and if it is found it will have a decided advantage over the clover as it can be grown in sour soil.

This little bean has a big botanical name, *Strophostyles helvola,* but the name means something. The generic part, *Strophostyles,* means twisted style; exactly what the plant's flower has. *Helvola,* the name of the species, means bay in color, referring to the flowers, of course. Britton, like Linnæus, could find sensible names for plants. The botanical name must needs be from the dead languages, but it should have meaning, and fortunately several of the world's distinguished botanists have known the classical languages well enough to use them properly. We could call this plant "the bay-flowered crooked-styled legume," and that is exactly what Britton called it when he constructed, from the dead languages, *Strophostyles helvola.*

We have so few good plants that are just weeds, this humble little wild bean vine comes like an angel among a host of sinners, a plant without a fault and yet just a weed.

TECHNICAL DESCRIPTION

Strophostyles Ell. Keel of the corolla with the included stamens and style elongated, strongly incurved, not spirally coiled. Pod linear, terete or flattish, straight or nearly so. Seeds quadrate or oblong with truncate ends, mealy-pubescent or glabrate; hilum linear.—Stems prostrate or climbing, more or less retrorsely hairly. Stipules and bracts striate. (Name from two Greek words meaning *a turning* and *style*.)

Strophostyles helvola (L.) Britton. *Annual;* stems branched, 0.3–2 m. long; leaflets ovate to oblong-ovate (rarely linear-oblong), *with a more or less prominent rounded lobe toward the base (the terminal 2-lobed),* or some or all often entire, 1.2–4 cm. long; corolla greenish-white and purplish; pod terete, 5–7.5 cm. wide, 4–8–seeded, nearly glabrous; *seeds oblong, about 6 mm. long,* usually very pubescent.—Sandy shores and riverbanks, coast of Massachusetts and southward; along the Great Lakes to Minnesota, and south to Kansas and Texas. June–September.

WILD GERANIUM OR CRANESBILL

[*Geranium Carolinianum* L.]

THERE are several wild geraniums, but the worst weed of the lot is the one known to the botanists as the Carolina Cranesbill. The word Geranium is from the Greek and means Cranesbill, so *Geranium Carolinianum,* the botanical name of the plant, is just the Latin way of saying Carolina Cranesbill. The long, sharp-pointed fruits of these plants are supposed to resemble the beaks of cranes.

Two or three of the Cranesbill species are very pretty plants. Their flowers are large enough to make a showing in a garden, but this particular species is not one of them. It has all of the requisite characteristics of a weed. It lacks beauty, it is worthless, it grows rapidly and in any place left uncultivated long enough for it to finish its quick growth, and it matures an abundance of seed before the owner of the place realizes that the squatter has moved in on his premises.

From one root many branches sprawl out in all directions. Every branch has many branches and these end in flower clusters whose inconspicuous flowers soon give place to clusters of beak-like fruits. The seeds at the lower ends of the beaks are dispersed by a quick splitting off of the sides of the beaks. The strips that split off let go first at the bottom where the seeds are attached and then curl upward, remaining attached at the top of the beak. As this abrupt splitting results in a jerk it pitches the seed to a considerable distance—a clever way to scatter seeds.

The seedling Cranesbill makes a pretty little rosette which persuades the unsuspecting gardener that here is some new flower he did not know he had. He leaves it a few days and returns to find a great sprawling weed already in bloom. Such is its weedy

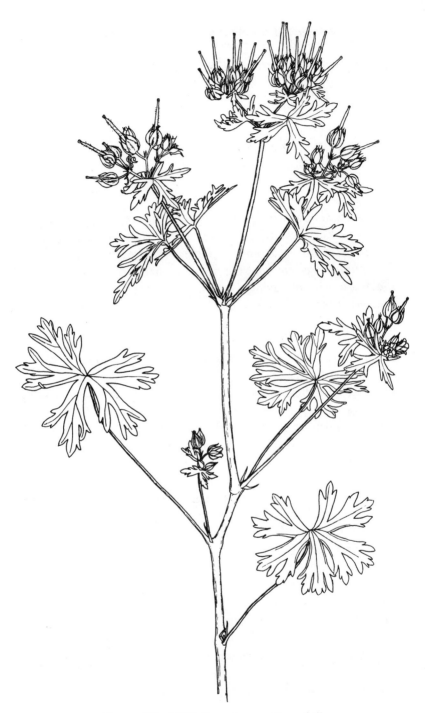

Fig. 43. The Wild Geranium or Cranesbill.

way: to gain space by deception and then to rush to fruition before the mistake is discovered. Most of these weeds will have completed their growth and scattered their seed before June 15 along the fortieth degree of latitude, and yet flowering and fruiting plants may be found throughout the summer if the season is not too dry.

The weed is at its worst in gardens, strawberry beds, and in sparsely set clover and alfalfa fields. In the well-kept gardens it serves as a fertilizer, but in the other places mentioned it crowds out and down the crop plants. It also does what all rapid-growing weeds are able to do: it filches from the soil the necessary plant foods intended for the crop plants.

TECHNICAL DESCRIPTION

Geranium (Tourn.) L. Stamens 10 (rarely 5), all with perfect anthers, the 5 longer with glands at their base (alternate with the petals). Styles smooth inside in fruit when they separate from the axis. —Stems forking. Peduncles 1–3–flowered. (An old Greek name from the word meaning a crane; the long fruit-bearing beak thought to resemble the bill of that bird.)

Geranium carolinianum L. Diffusely branched, hairy; leaves about 5–parted, the divisions cut and cleft into many oblong-linear segments; *flowers glomerate-cymose;* sepals ovate, about as long as the *whitish or very pale pink petals;* beak of fruit tipped with a *short filiform style;* seeds ovoid, minutely reticulated. Rocky places, etc., mostly in poor soil, eastern Massachusetts, southward and westward, common; May–June.

THREE-SEEDED MERCURY

[*Acalypha virginica* L.]

THERE is a very common little weed found in cultivated fields, in gardens and in pastures throughout the eastern half of the United States, and yet it is almost never named by those who notice it. If it is named it is called Three-seeded mercury, Wax-balls, Mercury weed and Copper leaf. The name, Mercury, is an old one. Pliny, a great Roman naturalist, called one of the spurges that looks somewhat like this one *Mercuriales,* dedicating the plant in this way to the god Mercury. The plants of this genus, *Acalypha,* resemble some of those of the genus *Mercuriales,* and so we might expect this one as well as those of the genus bearing the name to be called mercury. But there is another reason for calling it mercury. The bracts around the plant's seeds are shaped like the wings on the feet of the god Mercury. It may be that this is why it is called "Mercury weed" and "Three-seeded mercury." It has Mercury's wings, and it is three-seeded.

The name "Waxballs" refers to the waxy balls of pollen on the plant's stamens. They are rather conspicuous when the plant is in full bloom. Copper leaf is a good name as it is descriptive when the plant is mature.

One ought to be able to identify Three-seeded mercury by the names here given. The plant is often seen in company with the spiny sida. The two weeds are very much alike in size and shape, but their flowers are very different. The sida is a mallow with mallowlike flowers; the Mercury has the flowers of the spurges —Snow-on-the-mountain, Poinsettia, and such-like Euphorbias.

Three-seeded mercury is a weed worth knowing, not because it amounts to much, but because it is so very common and so sure to play a part in most of the gardens and cultivated fields east of the Rocky Mountains. The leaves of the plant look something like those of the nettles, persuading Linnæus to use the ancient name of the nettle, *Acalypha,* for the name of the genus to which this species, *virginica,* belongs. Here again Linnæus applied the name of the state to a species. He received specimens of the plant from that part of America known at that time as Virginia. Therefore we have *Acalypha virginica:* the nettle-leaved plant from Virginia.

TECHNICAL DESCRIPTION

Acalypha L. Flowers monœcious; the sterile very small, clustered in spikes; the few or solitary fertile flowers at the base of the same spikes, or sometimes in separate ones. Calyx of the sterile flowers 4–parted and

F<small>IG</small>. 44. Three-seeded Mercury. Note the foot-wings
of the god around the flowers

valvate in bud; of the fertile, 3–5-parted. Corolla none. Stamens 8–16; filament short, monadelphous at base; anther-cells separate, long, often worm-shaped, hanging from the apex of the filament. Styles 3, the upper face or stigmas cut-fringed (usually red). Capsule separating into 3 globular 2-valved carpels, rarely of only one carpel.—Herbs (ours annuals), or in the tropics often shrubs, resembling Nettles or Amaranths; the leaves alternate, petioled, with stipules. Clusters of sterile flowers and a minute bract; the fertile surrounded by a large and leaflike cut-lobed persistent bract. (*Acalypha* from an ancient Greek word meaning Nettle.)

Acalypha virginica L. Smoothish or hairy, 3–6 dm. high, often turning purple; *leaves ovate or oblong-ovate,* obtusely and sparsely serrate, long-petioled; *sterile spike* rather few-flowered, mostly *shorter than* the large *leaflike* palmately 5–9-cleft fruiting *bracts;* fertile flowers 1–3 in each axil.—Fields and open places, Nova Scotia to Ontario and Minnesota, south to the Gulf. July–September.

SNOW–ON–THE–MOUNTAIN

[*Euphorbia marginata* Pursh.]

SNOW-ON-THE-MOUNTAIN! The person who originated that name was a poet. It took imagination to see that the green and white on the top of these weeds resembled the ragged green-and-white strips extending down from a snow-crowned peak.

Snow-on-the-mountain is a beautiful weed; so beautiful that it is often used as a decorative plant in flower gardens and landscapes, and yet it is just a weed. It takes over pasture lands in the plains region of the West, and it is capable of doing the same thing to pastures east of the Mississippi River. It is more of a dry land plant, however, than one would expect it to be, and so we find it at its best out near where some of its cactuslike relatives would thrive. Several of the African species of Euphorbia are cactus plants.

The weed is an Euphorbia, which means that it is a plant with a "milky acrid juice," and that it is related to all of the spurges. The spurges include the crotons, the poinsettias, the crown of thorns, which is much like a cactus, the castor-oil plant, and all

FIG. 45. Snow-on-the-mountain. Note the seed
pods among the flowers

of the spurge weeds. Some of its relatives have medicinal properties, and most of them have the reputation of being poisonous. The Snow-on-the-mountain is undoubtedly poisonous, but it is of no medicinal worth.

There is little need of a description of the plant or even of an explanation of its botanical name. It is known to more people by its common name than almost any other plant treated in this book unless it be the dandelion, so there is little left to be said except to suggest a way to control or get rid of it.

Since the weed is more of a nuisance in pastures than anywhere else, the best method of control is to keep it from going to seed by mowing it down before the seed is set. If it comes into fields that can be cultivated it should be remembered that Snow-on-the-mountain makes just as good food for soil bacteria as any other plant, and that one can raise his soil fertility by turning under the weeds before the seed is set.

TECHNICAL DESCRIPTION

Euphorbia L. Flowers monœcious, included in a cup-shaped 4–5-lobed involucre (*flower* of older authors) resembling a calyx or corolla, and usually bearing large thick glands (with or without petal-like margins) at its sinuses. Sterile flowers numerous and lining the base of the involucre, each from the axil of a little bract, and consisting merely of a single stamen jointed on a pedicel like the filament; anther-cells globular, separate. Fertile flower solitary in the middle of the involucre, soon protruded on a long pedicel, consisting of a 3–lobed and 3–celled ovary with no calyx (or a mere vestige). Styles 3, each 2–cleft; the stigmas therefore 6. Pod separating into three 1–seeded carpels, which split elastically into 2 valves. Seed often caruncled.—Plants (ours essentially herbaceous) with a milky acrid juice. Peduncles terminal, often umbellate-clustered; in the first section mostly appearing lateral, but not really axillary. (Named for *Euphorbus,* physician to King Juba.)

Euphorbia marginata Pursh. Stem stout, 3–9 dm. high, erect, hairy; leaves sessile, ovate or oblong, acute; umbel with three dichotomous rays; glands of the involucre with broad white appendages.—Minnesota to Missouri, Colorado, Texas, and South Carolina; spreading eastward to Ohio, and frequently escaping from flower-gardens.

SPOTTED SPURGE

[*Euphorbia nutans* Lag.]

THERE are two spotted spurges. One of them lies flat, as flat as a carpet weed, and it has very small leaves with a spot on each of them. Its botanical name is *Euphorbia maculata,* meaning the spotted euphorbia. But the subject of this sketch is a larger-leafed plant that is somewhat erect. Its leaves are not always spotted. Many of the plants never have spots on them, but many of them do, and some of them have both spotted and unspotted leaves. It is a very common weed in gardens and cultivated fields after hoes and cultivators have ceased, which usually means after the spring-time gardening urge has cooled with the increase of atmospheric temperature. In other words, the spotted spurge, like pursley (or purslane), waits until the summer has fully arrived and then makes seeds while the sun shines.

This is an easy plant to identify and one with enough individuality to become possessed of several names. It is called Eye-bright, Nodding spurge, Milk purslane (a very descriptive name when one knows the purslane), Stubble spurge (it is often found in stubble fields after the small grain crops have been harvested), and Slobber-weed. The botanical name, *Euphorbia nutans* Lag. means the nodding Euphorbia named by Mariano Lagasca. The genus Euphorbia was named in honor of the famous Greek physician, Euphorbus. The plant is said to possess some medicinal properties, but none of its names, unless it be Eye-bright, refers to such characters. Its reddish cast, the milk in its stems and leaves, and the spots, when they appear, conspicuous dark patches, one on each leaf, are all identifying marks.

The weed is worthless so far as the husbandman is concerned, unless used to feed the soil bacteria. One always thinks of the plant as a companion of the pursley, but pursley can be eaten by man, and hogs thrive on it. Nothing eats spotted spurge. The spurges are all poisonous to a greater or less degree, and this one is fit only to hoe into the ground where it soon mixes forever

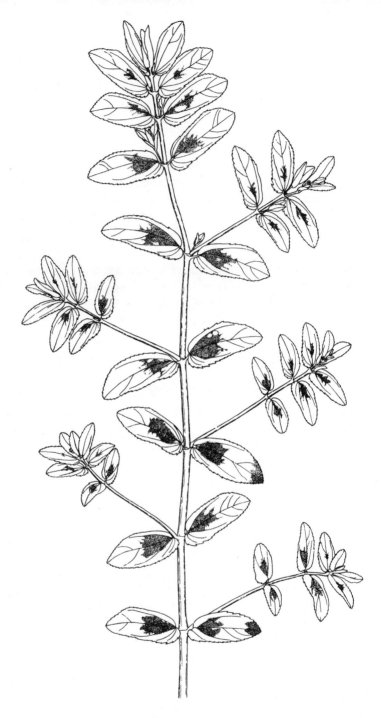

Fig. 46. The Spotted Spurge of gardens
and cultivated fields

with the elements, and where the cultivated plants soon send their roots abroad to pierce its mould.

TECHNICAL DESCRIPTION

See Snow-on-the-mountain for technical description of *Euphorbia* L. *Euphorbia nutans* Lag. Stem often subsimple below, erect or obliquely ascending, 2–10 dm. high; leaves oblique at the obtuse or slightly cordate base, ovate-oblong or oblong-linear, sometimes falcate, serrate, 1–3 cm. long, usually with a red spot or red margins; stipules triangular; peduncles longer than the petioles, collected in loose leafy terminal cymes; *appendages entire,* larger and white, or smaller and sometimes red; *pod glabrous; seeds ovate, obtusely angled, wrinkled and tubercled,* 1 mm. long, blackish. (*E. hypericifolia* Man. ed. 5.)—Dry open soil, Massachusetts to Ontario, Wisconsin, Nebraska, and southward.

POISON IVY

[*Rhus toxicodendron* L.]

WITH alliterations, we may almost truthfully say "the worst weed of the woods" or "the worst weed of waste places" is the poison ivy. If its toxin affected the skin of every one as it does that of a few it would not be necessary to qualify these alliterations. They would be literally true. This is the one weed, however, that every one who lives in the country and all who visit it should know. Those who are immune to its poison have nothing to fear, but no one knows whether he is immune or not until he has come in contact with some part of the plant. If he is susceptible he will long remember that day, and if he is wise he will develop his observational powers until they equal those of a first-class botanist when he approaches a likely habitat of this snake of the weeds.

The poison ivy is strictly American, and is found in nearly every part of the United States and Canada. It is sometimes called Poison oak, Poison vine, Poison creeper, Three-leaved ivy, Picry, and Mercury. The plant takes many different forms and botanists have been led to give to some of these forms specific names, but many of the best taxonomists hold that the so-called species are merely varieties and that whether it stands or crawls it is the same old Satan and should be designated as such.

By its leaves you should know it, not by its fruits, and the best way to know its leaves is to see enough of them to completely saturate the memory. There are no leaves exactly like those of the poison ivy, of course, but there are many leaves made up in the same way. The name, "Three-leaved ivy," attempts to describe the leaves, but it actually does not. The plant is not three-leaved. Its leaves are compound and are so made up with three leaflets to each leaf. Each leaf is attached to the stem of the plant with a long leaf-stem (petiole) and the petiole carries three leaflets that are characteristically crinkled and with round-toothed or slightly lobed edges. It is in the leaves that the most variation is seen, and, strikingly enough, the more one looks for variation the more variation he can see, until he begins to think that no two of these hateful vines are alike, and that he has to learn his leaves all over again whenever he unguardedly rushes in where angels would not. With patience, however, the leaves come to have a general appearance that no others possess, and the student of this one plant has then developed what the botanist has for all plants: the ability to see differences and similarities at a glance.

The botanical name, *Rhus toxicodendron,* is intended to give the nature of the weed. *Rhus* is the ancient name for sumac both in the Greek and the Latin. Poison ivy is a sumac but not *the* poison sumac. That is a cousin, *Vernix* by name, *Rhus vernix.* It contains as bad blood (bad juice) as the cousin we are treating, but it is not so often met with.

The *Toxicodendron* part of the name is a combination of two Greek words, *toxico* meaning poisonous and *dendron* meaning tree or plant. The word, therefore, means the toxin-bearing plant.

Poison ivy is not a hard weed to eradicate. It seldom comes back after it has been pulled up; and any one who knows himself immune to its toxin can easily tear it from its rather insecure anchorage with his bare hands. It grows readily from seed, and the few birds that eat its fruit (little white berries) scatter the seeds in their excrement. The downy woodpecker is said to be one

FIG. 47. The Poison or Three-leaved Ivy

of the worst offenders, but let him who is without sin throw stones or shoot lead at the downy. Scattering poison ivy seeds is his only fault and his life's processes are served by that fault.

He who is poisoned by the weed wants to know what he can do about it. Here are two remedies that usually effect a cure. If one of these fails after twenty-four hours of treatment, try the other one.

Squeeze the juice of the wild touch-me-not (see page 155) on the affected skin, or rub the bruised leaves and stems over the skin until it is wet with the juice. This is an absolute cure for some people. A young lady who suffered intensely from ivy poisoning was able to stop it not only with the juice of the wild touch-me-not, but even by soaking the withered stems and leaves of the plant in water and then applying the soaked plant tissues to the affected skin.

Another remedy for most people is the painting of the poisoned skin with a 5 per cent aqueous solution of potassium permanganate. This will color the skin an ugly brown, but the color can easily be removed after the cure by applying to it a weak solution of oxalic acid.

If absolute alcohol is available it is declared to be an absolute cure for ivy poisoning.

TECHNICAL DESCRIPTION

Rhus L. Calyx small, 5-parted. Petals 5. Stamens 5, inserted under the edge or between the lobes of a flattened disk in the bottom of the calyx. Fruit small and indehiscent, a sort of dry drupe.—Leaves usually compound. Flowers greenish-white or yellowish. (The old Greek and Latin name.)

Rhus toxicodendron L. Suberect and bushy, scrambling over fences, walls, etc., or in woods climbing by rootlets to considerable heights, sparingly pubescent or glabrate; *leaves pinnately 3-foliate, leaflets ovate to rhombic, mostly acuminate, entire, crenulate, or irregularly and coarsely few-toothed,* paler and with some persistent or tardily deciduous pubescence beneath; *berries* whitish or cream-colored, subglobose, *glabrous or nearly so, 5-6 mm. in diameter,* in age sulcate.—Abundant in hedgerows, thickets, and woods. June–July.—Passing on our western limits to a thicker-leaved smoother form.

WILD TOUCH-ME-NOT

[Impatiens biflora Walt; *Impatiens palida* Nutt.]*

THERE is but one reason for including the wild touch-me-not in this book of weeds: it is the weed every one susceptible to poison ivy should know. Rubbed on the affected skin the juice of this plant frequently brings almost immediate relief, and a complete cure is likely to result in twenty-four hours.

Wild touch-me-not grows in wet, shady places, especially along little rills; either dry or running rills. There are two species: one with orange-colored flowers and one whose flowers are rather pale yellow, but both species are alike in every way except in the color of their flowers. The orange-colored one is called the Spotted touch-me-not, the other is generally called the Pale touch-me-not. The "touch-me-not" refers to the way the seed pod behaves when it is touched. It impatiently flies to pieces and curls up.

The stems of the plants are almost glassy in appearance. They are found when crushed to be gorged with a watery juice, and the peculiar swellings at the nodes of the stems seem to be reservoirs for the storage of this juice. The plants can easily be identified, even before they bloom, by their leaves and by their smooth, almost transparent stems with swellings at the nodes.

It is useless to look for the weed except where it grows. If one knows a shaded little glen, perhaps shaded by only a few shrubs and a tree or two, a glen made rich by the washed-in dirt from the fields above, he is almost sure to find the jewel weed (another name for the wild touch-me-not) growing there along the little ditch made by the spring rains. It seldom grows by a running stream unless it be a mere rill from a spring. By such a rill, if there is shade enough, one can expect to find either one or both of the species growing in abundance.

If you are susceptible to the poison of the ivy you should spot a natural habitat of the wild touch-me-not before the growing season starts. Watch the place and if there appears a host of weeds that are "glaucous, succulent annuals" and if the leaves are ovate with petioles and if the stems seem filled with a watery juice, shout "Eureka," and go back there for relief whenever you find yourself "breaking out" with the poison pimples. Of course you may find the plant ineffective in your case, but it may be the very remedy you require, and if it is, you will be forever grateful for the quirk that placed this harmless plant among the hundred and two weeds treated and illustrated in *Just Weeds.*

The botanical name, *Impatiens,* is from the Latin and refers to the impatient, touch-me-not character of the seed pod. The two specific names, *biflora* and *pallida,* are self-explanatory to him who sees the plants. The orange-colored flowers are usually in twos, therefore, *biflora.* The pallid color of *pallida* explains its name.

Some of the common names for the weed are Jewel-weed, Silver cap, Wild balsam, Lady's ear drops, Snap weed, and Wild lady's slipper.

TECHNICAL DESCRIPTION

Impatiens (Rivinius) L. Sepals apparently only 4; the anterior one notched at the apex (probably two combined); the posterior one (appearing anterior as the flower hangs on its stalk) largest, and forming a usually spurred sac. Petals 2, 2–lobed (each a pair united). Filaments appendaged with a scale on the inner side, the 5 scales connivent over the stigma; anthers introrse. Pod with evanescent partitions, and a thick axis bearing several anatropous seeds; valves 5, coiling elastically and projecting the seeds in dehiscence.—Leaves in ours ovate or oval, coarsely toothed, petioled. Flowers axillary or panicled, often of two sorts, viz., the larger ones which seldom ripen seeds; and very small ones which are fertilized early in the bud, their floral envelopes never expanding but forced off by the growing pod and carried upward on its apex. (Name from the sudden bursting of the pods when touched, whence also the popular appellation.)

Impatiens biflora Walt. *Flowers orange-colored, thickly spotted with reddish brown; sac longer than broad,* acutely conical, tapering into a *strongly inflexed spur* half as long as the sac. (*I. fulva* Nutt.)—Rills and shady moist places. June–September.—Plant 6–8 dm. high. Forms white; spotless, whitish, or roseate flowers have been found.

Fig. 48. Wild Touch-me-not, a cure
for poison ivy

VELVET-LEAF

[*Abutilon theophrasti* Medic.]

VELVET-LEAF, velvet-weed, Indian mallow, butter print, pie print, Cotton-weed, and Indian hemp are all rather descriptive names of this common weed found in cornfields throughout the Mississippi Valley. The weed has velvety leaves, and its seed pod looks very much like the print block used by farm housewives for stamping their rolls of butter. The name, Indian hemp, may not seem so descriptive to him who has never broken one of the dead weeds. It is then that the long, strong fibers of the bark are seen; such fibers as might well be used for making thread or cord. The botanical name is not descriptive in the least. The origin of the name, *Abutilon,* is unknown, and the specific name, *Theophrasti,* is given in honor of the ancient naturalist, Theophrastus.

Velvet-leaf weeds range in height from two to six feet. They have big, heart-shaped, velvety leaves that in themselves are enough to identify the plants, but the yellow, five-parted flowers and the butter-print seed pods make their identification certain.

The weed is a hot-weather plant and so does not grow much until hot weather comes, which means until the corn is laid by. In fact many of the seeds do not sprout before the middle of July. Like most weeds it wastes no time in growing after the seed sprouts, and it is for this reason one is often surprised to find the cornfields of some of the best farmers filled with the velvet-leaf. This fast-growing annual is thus able to escape the plow of the most painstaking farmer. The weeds were not there when the last plowing was done, but they were there soon after that, and but a few days later the flowers and seed pods appeared.

Like the cockle bur the velvet-leaf can grow almost anywhere, but it is at its best in the best soil. There it can vie with the strongest of cultivated plants, and corn, or any other crop, in a field filled with such weeds has to struggle against serious competition. Big weeds can get the water and the minerals needed

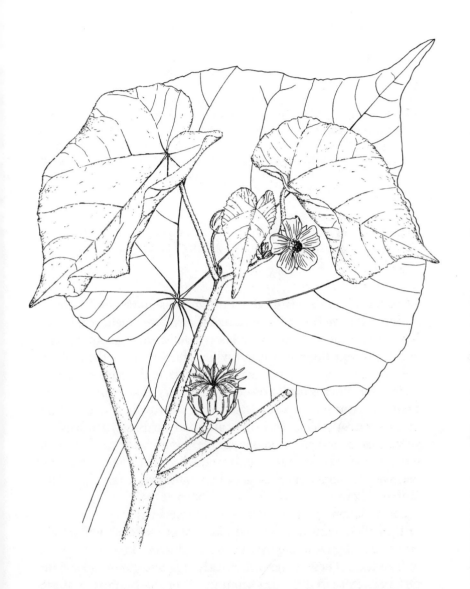

Fɪɢ. 49. Velvet Weed or Pie Print. Note the seed-pod printing-stamp

for their growth in spite of the crop plants. The velvet-leaf is a thief in this respect, and every grower of crop plants should know it.

TECHNICAL DESCRIPTION

Abutilon (Tourn.) Mill. Carpels 2–9-seeded, at length 2–valved. Radicle ascending or pointing inward. Otherwise as in *Sida*. (Name of unknown origin.)

Abutilon theophrasti Medic. Tall annual, 6–12 dm. high; leaves roundish-heart-shaped, taper-pointed, velvety; peduncles shorter than the leaf-stalks; corolla yellow; carpels 12–15, hairy, beaked.—Waste places, vacant lots in cities, etc. (Naturalized from India.)

FALSE MALLOW

[*Sida spinosa* L.]

WHEN one hoes in his garden during the summer he is almost sure to see a small weed that bears little, yellow, mallow flowers: flowers like those of the hollyhock, and with it another weed that looks something like it but which does not have the mallow flowers. The one with the mallow flowers is the subject of this sketch; the other is the Three-seeded mercury. The two weeds are about the same size, have the same habit of growth, and are found in like places; which means in cultivated fields and gardens, in waste places, and in worn-out meadows and pastures. The two weeds are not remotely related to each other. The one is a mallow, the other a spurge, but they are often confused; a fact that demonstrates how faulty casual observation may be.

False mallow is of the genus *Sida*: one of the genera of the great mallow family. The weed has at least five common names, two of which are descriptive when the observer knows the plant to be a Sida. These names are Prickly sida and Spiny sida. Both make reference to the rather blunt spines or sharp tubercles at the base of each leaf and branch of the plant. The spines are not true spines, but when one with his bare hands pulls up the wiry little weed he is made aware of the appropriateness of the names.

FIG. 50. The False Mallow is the Thorny Sida

Linnæus evidently thought the tubercles were spiny enough to be used in the descriptive scientific name he gave the weed. The other common names are False mallow, Indian mallow, and Thistle mallow.

The word *sida* is the Latin form of the Greek word *side,* which simply meant plant. Linnæus used the word because Theophrastus, a Greek philosopher and naturalist, had used it. It so happens that this little weed is one of four species found in this country. They are all weeds, but the genus to which they belong is a big one and several valuable tropical plants are Sidas. Queensland hemp is one of them. This plant—a first cousin of our little false mallow—yields a very fine fiber, said to be superior to jute in strength, and a valuable demulcent is derived from this plant, also. Several other tropical sidas are medicinal in nature and produce demulcents.

So our little insignificant weed, so insignificant as to go unnoticed except by botanists and by those who hoe in gardens, has some very important relatives. It stands out in contrast with the aristocrats of the genus like a wizened little tramp who rears back and declares that he is one of the great Sidas. Its very impudence gives it a place among *Just Weeds.*

TECHNICAL DESCRIPTION

Sida L. Calyx naked at the base, 5-cleft. Petals entire, usually oblique. Styles 5 or more, tipped with capitate stigmas; the ripe fruit separating into as many 1-seeded carpels, which are closed, or commonly 2-valved at the top, and tardily separate from the axis. Seed pendulous. Embryo abruptly bent; the radicle pointing upward. (A name used by Theophrastus.)

Sida spinosa L. Annual weed, minutely and softly pubescent, low (2.5–5 dm. high), much branched; *leaves ovate-lanceolate or oblong,* serrate, rather long-petioled; peduncles axillary, 1-flowered, shorter than the petiole; *flowers yellow,* small; *carpels* 5, combined into an ovoid fruit, *each splitting at the top into 2 beaks.*—Waste places, Massachusetts to Michigan, Kansas, and southward, where common.—A little tubercle at the base of the leaves on the stronger plants gives the specific name, but it cannot be called a spine. (Naturalized from the Tropics.)

THE COMMON MALLOW

[*Malva rotundifolia* L.]

IF THERE is such a thing as an innocuous weed, the common mallow comes near being it, but it has at least one weedy character, and that is the filling of space which might well be occupied by some useful plant.

The weed belongs to an interesting family since such plants as cotton, hollyhocks, and the Rose of Sharon belong to it, as do also those real weeds, velvet-leaf and the flower-of-an-hour. The family is recognized by the tube of many stamens that surrounds the pistil of each flower, and the genus to which the true mallows belong, *Malva,* is known by the ring of seeds found at the center of the floral parts after the petals have fallen. In the common mallow this ring of seeds looks like a little flat cheese, and so one of the plant's common names is "Cheeses."

The round leaves and the prostrate stems with their little flowers and seed rings make the identification of the common mallow easy. Any one who sees a prostrate plant making a loose circle eighteen inches to two feet in diameter, its round leaves the size of a dollar loosely covering the loose stems, may well guess that he is looking at the common mallow. Then if he sees the little five-parted flowers, and farther down on the stems he finds where the petals have fallen and left the purselike containers made by the folded-in sepals of the calyx, he can be sure he has found the common mallow, if in those purses he finds the "cheeses," the seed rings. The cheeses, if fresh, that is, still soft, are not unpalatable, and many a country boy has nibbled them.

The long root of the plant tastes somewhat like ginseng, and probably has just as much medicinal virtue as the ginseng has. The Greek name, *Malva,* refers to the salve-making properties and Mallow has been derived from Malva, but it has come to have the connotation of softness. Whether it is the soft leaves or the soft immature seeds the name refers to makes little difference, for the mallows have both. *Rotundifolia* means round leaves.

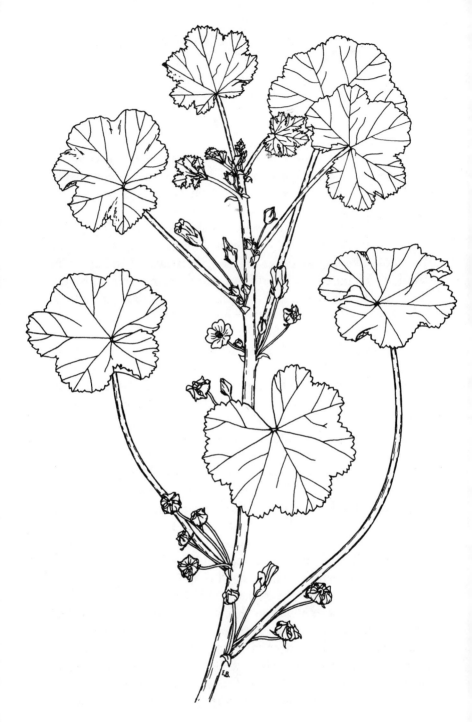

FIG. 51. The Common Mallow. It is often called Cheeses

TECHNICAL DESCRIPTION

Malva (Tourn.) L. Calyx with a 3-leaved involucel at the base, like an outer calyx. Petals obcordate. Styles numerous, stigmatic down the inner side. Fruit depressed, separating at maturity into as many 1-seeded and indehiscent round kidney-shaped blunt carpels as there are styles. Radicle pointing downward. (An old Latin name, from the Greek name having allusion to the emollient leaves.) *Malva rotundifolia* L. *Stems procumbent* from a deep biennial root; *leaves* round-heart-shaped, on very long petioles, crenate, *obscurely lobed;* petals twice the length of the calyx, whitish; carpels pubescent, even.—Waysides and cultivated grounds, common. (Naturalized from Europe.)

FLOWER-OF-AN-HOUR

[Hibiscus trionum L.]

THE Flower-of-an-hour is a beautiful weed. It is so beautiful that it was once used in flower gardens, from which it escaped in indignation when it was cast aside for flowers not so ephemeral. And although it now has to grow and bloom in cultivated fields and vegetable gardens scattered over the greater part of the eastern half of the United States it is still a beautiful and never a very bad weed. Its greatest fault is in the use of the plant food materials that might serve cultivated plants, but when that loss to agriculture is compared with the amount leached out and washed away by even the smallest of rains the Flower-of-an-hour has almost nothing to condemn it to the rogues' gallery of weeds.

The plant is a mallow, and so belongs with the hollyhocks (far worse weeds than it is), with the Rose of Sharon, the beautiful Marsh Mallows, and other hibiscus plants. It is also a relative of the prickly sida, the velvet-leaf and cheeses, three other mallow weeds treated in this book, but it is in none of their classes as a weed. In fact it would still deserve a place in flower gardens if it would only hold that flower for more than an hour.

The reader may know it by some of its other names: Bladder

Fig. 52. The Flower-of-an-hour, a relative
of the hollyhock

ketmia, Ketmia, Venice's mallow, Modesty, Shoo-fly, etc. It was imported from Europe and some of its names came with it. The name, Flower-of-an-hour, is the most descriptive even if it is the most cumbersome. The name is so descriptive that any one knowing the flower of the mallows would soon discover that this little plant with its deeply cut, palmate leaves and with its nearly transparent, bladderlike seed pods formed from the calyxes—any one seeing such a plant would soon discover that this was the flower-of-an-hour.

The botanical name, *Hibiscus,* is an old Greek and Latin name which seems to have no special meaning. *Trionium,* the specific name, refers to the three-parted leaf.

TECHNICAL DESCRIPTION

Hibiscus L. Calyx involucellate at the base by a row of numerous bractlets, 5-cleft. Column of stamens long, bearing anthers for much of its length. Styles united, stigmas 5, capitate. Fruit a 5-celled loculicidal pod. Seeds several or many in each cell.—Herbs or shrubs, usually with large and showy flowers. (An old Greek and Latin name of unknown meaning.)

Hibiscus trionum L. A low rather *hairy annual;* upper leaves 3-parted, with lanceolate divisions, the middle one much the longest; fruiting *calyx inflated, membranaceous, 5-winged, with numerous dark ciliate nerves;* corolla sulphur-yellow, with a blackish eye, ephemeral.— Cultivated and waste ground, rather local. (Naturalized from Europe.)

EVENING PRIMROSE

[Œnothera biennis L.]

THE MOST outstanding character of the evening primrose is its flower, which is designed to insure promiscuous mating. The plant babies which develop in the ovaries of the evening primrose are test-tube babies. Of course this same fact is true of many other plants, but the evening primrose is unique in its choice of the doctors required to bring the male principle (pollen grains) to the amorously waiting female organs. They are a selected lot: the sphynx moths, those night-flying insects that look like and behave like humming birds. They are just as ignorant as the sphynx itself of the work they are doing. All the moth doctors know is that there is a sip of nectar at the bottom of those long flower tubes. In getting that nectar their proboscises become dusted with the pollen grains of the flowers whose pollen grains happen to be ripe. Then down into another flower tube a proboscis goes, leaving some of that pollen dust on an anxiously waiting stigma, the part of the female organ whose business it is to receive and cause to develop the male sex cells that are contained in the pollen grains.

The processes that follow this simple act are very complex, just as they are in all sexual reproduction, but the result is the same: young are formed. They are primrose seeds, of course, and they are all wrapped up in a supply of food that will feed them until they have fastened their roots in the good earth and spread their tiny seed leaves to the glorious sun.

There are a great many evening primroses and several of them are weeds, but the one the botanists call *biennis,* meaning the biennial one, is among the weediest of the lot. It is a very common, rather coarse, erect weed that is seen in pastures and fields left fallow for a year or more. If branched, as it often is, its branches are as coarse as its stem. It may attain a height of six feet and it is seldom less than four feet. It is, therefore, a conspicuous weed, and especially conspicuous with its yellow flowers which open in the evening and do not close before nine o'clock the next day; and yet few of our farming population have a name for it. It has several names, however, such as Night willow-herb, Scabish, Tree primrose, Cure all, Field primrose, Fever plant, Coffee plant, and Wild beet. The last name refers to the beetlike appearance of the thick root of the plant during the first year of its growth. These roots have been used as food and as a source of home remedies. The rosettes of first-year leaves are used as greens (pot herbs) and are considered by some cooks as a very necessary constituent of that savory dish.

The weed is easily identified when it is in bloom. Any plant

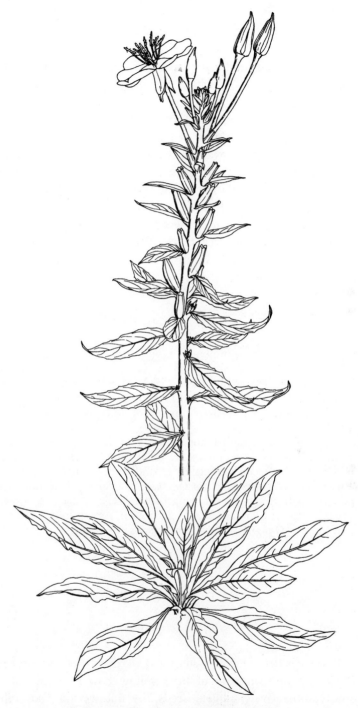

Fig. 53. The Evening Primrose with
its edible rosette below

from four to six feet tall that blooms out in the evening with only a few yellow flowers which become fragrant as the night comes on is surely the evening primrose. The rosettes are easily identified, also. The lanceolate leaves that stand up rather than lie down and that usually show red spots which look as if a dark wine had been sprinkled over them distinguish them from those of all other plants. The rosettes are worth knowing for the sake of greens. The mature plants do not amount to much except as sphynx-moth feeders, and surely we could get along without sphynx moths. The fact that the weed is a biennial keeps it from being aggressive in cultivated fields. Where it becomes troublesome enough to require eradication it should be plowed under and made to serve as a fertilizer.

TECHNICAL DESCRIPTION

Œnothera L. Calyx-tube prolonged beyond the ovary, deciduous; the lobes 4, reflexed. Petals 4. Stamens 8; anthers mostly linear and versatile. Capsule 4-valved, many-seeded. Seeds naked or with an obscure membranaceous crest.—Leaves alternate or rarely all basal. Flowers yellow, white or rose-color. (An old name of unknown origin, for a species of *Epilobium*.)

Œnothera biennis L. Rather stout, erect, 3–15 dm. high, usually simple, more or less spreading-pubescent to hirsute; leaves lanceolate to oblong—or rarely ovate-lanceolate, repandly denticulate, acute or acuminate; *bracts* lanceolate, *shorter than* or scarcely exceeding the *capsules;* calyx-tube 2.5–3.5 cm. long; petals yellow, obovate, 1.5–2.5 cm. long; *pods* more or less *hirsute,* narrowed almost from the base, 2–3.5 cm. long. (*Onagra* Scop.)—Open places, common.

WILD PARSNIP

[*Pastinaca sativa* L.]

WILD PARSNIP is the big, coarse-stemmed weed with perfect umbels (really compound umbels) of little yellow flowers, which soon become comparatively bigger "seeds," seen towering above the

FIG. 54. Wild Parsnip, a garden vegetable run wild

shorter weeds along highways and in waste places. It is the garden parsnip run wild, wild and dangerous. Strange as it may seem, this is the same plant whose roots in the first year of their growth are eaten and relished by some people. This coarse, fluted, and hollow stalk is sent up from one of those food-filled roots in the second year, and then the whole plant becomes poisonous.

Even in the first year of its growth the wet leaves are likely to irritate the skin, and these same leaves have been mistakenly gathered and eaten as greens (pot herbs) with rather serious results in some instances.

So the Wild parsnip is a big, bad weed, but one easy to eradicate. If all the wild parsnips in the world were cut down in the blooming stage there would be no more wild parsnips; unless some one left his parsnip roots in the ground until they seeded, or until an escape was made from a seedman's fields. Anyway, all one has to do to rid himself of this ugly, worthless, dangerous weed is to mow it down before the seeds are set.

In spite of the fact that every one who knows the plant seems to know it as the Wild parsnip, it has at least three other common names: Bird's-nest, Hart's-eye, and Madnip. Perhaps it is to be expected that outlaws have aliases; and perhaps those aliases should not describe the outlaw. Two of these certainly do not describe the Wild parsnip, and Bird's-nest, the only one of the three that is remotely descriptive, should be reserved for the Queen Anne's lace.

The genus to which the Wild parsnip belongs is a very small one. Its name, *Pastinaca*, is Latin for parsnip and is derived from the Latin word *pastus*, meaning food. The specific name, *sativa*, is given to plants that are grown for food. It really means a cultivated plant.

There are some interesting facts given about the Wild parsnip in one of the Herbals. It is said that in Ireland beer is made from the roots of the first year's growth. A good wine is made from them in some other countries, and in other places by distillation a sort of rum is made from these selfsame roots. Perhaps the weed has a reason for being after all.

TECHNICAL DESCRIPTION

Pastinaca L. Calyx-teeth obsolete. Fruit oval, very much flattened dorsally; dorsal ribs filiform, the lateral extended into broad wings, which are strongly nerved toward the outer margin; oil-tubes small, solitary in the intervals, 2–4 on the commissure; stylo-podium depressed. —Tall stout glabrous biennial, with pinnately compound leaves, mostly no involucre or involucels, and yellow flowers. (The Latin name, from *pastus,* food.)

Pastinaca sativa L. Stem grooved; leaflets ovate to oblong, cut-toothed. —Waste places, open rich soil, etc. (Naturalized from Europe.)

QUEEN ANNE'S LACE

[*Daucus carota* L.]

ONE OF the worst weeds that has come to the United States from Europe is Queen Anne's Lace. It is also called Wild Carrot (which it really is), Bird's-nest, Crow's-nest, and Devil's plague—another thing it really is.

There is no doubt that the weed is a wild carrot; authorities say that it is the cultivated carrot run wild. If it is it has run a long way and into all sorts of places. It seems never to be the least particular in its choice of a habitat. It thrives in uncultivated ground wherever or whatever it may be so long as it is not densely shaded.

The name, Queen Anne's Lace, is so descriptive of the flower-cluster that it, rather than any of the other names, is selected for this sketch. After the flowers are gone the umbel with its developing seeds draws together in such a way as to form the "bird's nest" which gives to the plant another of its descriptive names.

The umbel of flowers is the weed's most striking feature, however, and it is not only its striking resemblance to lace that makes it so. At the very center of this umbel of snow-white flowers is a single purple, almost black, flower. Like all the other flowers

FIG. 55. Queen Anne's Lace or Bird's Nest

of the umbel it is very small, but what a contrast! What can be its purpose? What possessed old Mother Nature to attempt to hide that purple stitch in the center of each of the Queen's snow-white kerchiefs? Well, it is certainly an identification mark. Any one who sees the weed for the first time knows what it is when he finds that little, dark-purple flower at the center of the lacy flower cluster.

Queen Anne's Lace is a medicinal plant. Its extract or tea may be used as a stimulant, as a diuretic (it acts on the kidneys), and as a deobstruent, which means as an aperient, and that means something in the way of a laxative. The plant contains a volatile oil that smells like turpentine. The "birds' nests" seem to be saturated with this oil just before the seeds ripen.

As a weed Queen Anne's Lace has few equals. It can take over meadow and pasture lands with the greatest of ease, for no pasturing stock will touch it and it is tall enough to crowd out nearly all of the grasses used for hay. When it is once established in a field there is only one sensible way to rid the field of it and that is to use it as a fertilizer, plowing it under just before it reaches the blooming stage. If this is done, and if the field is kept in some cultivated crop for two or three years thereafter, the weed will be whipped in that particular field, but of course it will still be hanging around the edges. It scatters its seeds from roadsides and fence rows where neglect and the weed hold sway. It is not possible to use it as a fertilizer in such places, but it is possible to keep it from seeding there, and the few hours required to mow clean all fence rows and roadside strips where Queen Anne's Lace is growing are hours well spent.

The generic name, *Daucus,* is a modification of the Greek name of the plant. *Carota,* the specific name, is Celtic and means red of color; so *Daucus carota* L. means the red-rooted umbeliferous plant named by Linnæus.

TECHNICAL DESCRIPTION

Daucus (Tourn.) L. Fruit oblong, flattened dorsally; stylopodium depressed; carpel with 5 slender bristly primary ribs and 4 winged secondary ones, each of the latter bearing a single row of barbed prickles; oil-tubes solitary under the secondary ribs, two on the commisural side.—Bristly annuals or biennials, with pinnately decompound leaves, foliacious and cleft involucral bracts, and compound umbels which become strongly concaved. (The ancient Greek name.)

Daucus carota L. Biennial; stem bristly; ultimate leaf-segments lanceolate and cuspidate; rays numerous.—Fields and waste places; a pernicious weed.—The flowers vary from white to roseate or pale yellow, the central one in each umbel usually dark purple. (Naturalized from Europe.)

MONEYWORT

[*Lysimachia nummularia* L.]

MONEYWORT is a beautiful little creeping plant that may become a serious pest where shade and moisture favor it. It can be used as a decorative plant on banks and in rock gardens if the banks and the rock gardens are not too dry. It becomes a weed in lawns and in pastures but never in cultivated grounds.

The plant gets its name, moneywort or money plant, because of its coin-shaped leaves, but it has many other names. Some of them are, Creeping Loosestrife, Creeping Jenny, Creeping Joan, Wandering Taylor, Herb twopence, Twopenny grass, Yellow Myrtle, and Creeping Charlie. Such is the way of the English, and such is the way of common names. Every locality seems to have originated one or more names for this lowly weed.

And believe it or not, this innocent little thing is among the medicinal plants. It was, and perhaps still is, believed to be useful as a blood stanch. There is a superstition connected with it to the effect that wounded snakes crawl over and lie on moneywort leaves to heal their wounds. It was at one time called *Serpentaria* because of this belief. The bruised leaves are said to be classed as a subastringent.

The moneywort is so easily identified that a description of it is scarcely necessary. There is not another plant like it: a creeping little vine that takes root every little way; leaves opposite, round

FIG. 56. The Moneywort; they say it
cures wounded snakes

as coins and with almost no stems to them, flowers yellow and conspicuous; that is the moneywort.

The generic name, *Lysimachia,* may be in honor of King Lysimachus or it may be made up from two Greek words, one meaning *a release from,* and the other meaning *strife.* If the latter is the derivation then loosestrife is a good name for the genus. The specific name, *nummularia,* means coin. So *Lysimachia nummularia* L. means the loosestrife with coin-shaped leaves named by Linnæus.

TECHNICAL DESCRIPTION

Lysimachia (Tourn.) L. Calyx 5–6–parted. Corolla rotate, the divisions entire, convolute in bud. Filaments commonly monadelphous at base; anthers oblong or oval. Capsule few-several-seeded. Leafy-stemmed perennials, with herbage commonly glandular-dotted. (In honor of King *Lysimachus,* or from two Greek words, one meaning *a release from,* the other *strife.*)

Lysimachia nummularia L. Smooth; stems trailing and creeping; leaves roundish, small, short-petioled; divisions of the corolla broadly ovate, obtuse, longer than the lance-ovate calyx-lobes and stamens. Escaped from gardens into damp ground in some places. (Introduced from Europe.)

DOGBANE

[*Apocynum cannabinum* L.]

NEARLY every one who sees this weed and has a little learning thinks that he is looking at a milkweed. The weed does have milk in it and its pods, somewhat like those of the milkweed, pop open and allow the wind to carry away the downy seeds, which, again, are very like those of the milkweeds. But in spite of all of its milkweedy characters it is not a milkweed. It is Dogbane, Indian hemp, American hemp, Indian physic, Choctaw root, Bowman's root, Rheumatism weed, and Hemp dogbane, but never milkweed.

Fig. 57. The Dogbane that many call a milkweed

There are two reasons why this plant is not a milkweed even though it is a weed filled with milk. The first reason is that it has its pods in pairs, and second, its flowers are like the flowers of other plants, not made for fooling and trapping poor little insects. That is what the milkweed flowers do, whether the weeds have milk in them or not; and some milkweeds are like some cows: milkless. They all have the milkweed flower, however, a flower that cannot be described here, but suffice it to say that the bees and butterflies, insects that know all about flowers, are fooled by the flowers of the milkweeds. Milkweeds have single pods also.

So the dogbane, with all its milk, is no more a milkweed than a goat is a cow. It may deserve the name, but it deserves even more to be called Indian hemp, since its fibrous bark was used by the Indians to make cords and cloth. Any one who knows the plant can prove to his own satisfaction the fibrous nature of the bark by breaking a mature stalk. Some birds gather the fibers from old stems for their nest building.

The roots of the weed were used by the Indians as medicine, and for this reason "Indian physic" and "Choctaw root" are applicable names. Tincture Apocyne comes from this plant, and so it has a place in some of the issues of the United States Pharmacopœias and in the United States Dispensatory.

As a weed it is not very aggressive, but like all perennials it is persistent. It likes sour, rich ground, and so is often seen in abundance in waste places near industrial plants. There it has little competition, and if the soil is rich enough it will attain in such places its maximum height of nearly six feet. Its usual height is not more than three feet. The smooth, reddish stems and the long slender pods in pairs will identify the weed. Look for it in waste places.

TECHNICAL DESCRIPTION

Apocynum (Tourn.) L. Calyx-lobes acute. Corolla bell-shaped, bearing 5 triangular appendages below the throat opposite the lobes. Stamens on the very base of the corolla; filaments shorter than the arrow-shaped convergent anthers, which slightly adhere to the stigma. Style none; stigma large, ovoid, slightly 2-lobed. Fruit of 2 long and slender follicles. Seeds with a tuft of long silky down at the apex.—Perennial herbs, with upright branching stems, opposite mucronate-pointed leaves, a tough fibrous bark, and small and pale cymose flowers on short pedicels. (Ancient name of the dogbane composed of two Greek words, the first meaning *from,* the second *dog.*)

Apocynum cannabinum L. Glabrous, 2-24 dm. high, the stems and branches ascending (but on gravel beaches, etc., depressed and wide-spreading); *leaves* mostly ascending, *usually pale green,* ovate-oblong to lanceolate, *glabrous or sparingly pubescent beneath, those of the chief axis narrowed at base to distinct petioles* (2-7 mm. long), *those of the branches often subsessile; central cyme flowering first; flowers erect;* calyx glabrous, its lobes about equaling the corolla-tube.—Gravelly or sandy soil, mostly near streams; on beaches becoming dwarfed and diffuse, with smaller and narrower leaves. (*A. album* Greene.) June–August. Varying greatly.

(The specific name *cannabinum* comes from the Latin name for hemp.)

MILKWEED

[*Asclepias syriaca* L.]

IT IS not the milk that makes the milkweed but the striking flower it has. It is true that the most of the plants in the family, *Asclepiadaceæ,* have a milky juice, but there are a few of them that are about as devoid of the lactic fluid as a Texas steer. Whether they have milk or not they have one of the most distinctive flowers in the plant kingdom. It is made for the purpose of tricking insects in order that cross fertilization may be assured. The mechanism is so unique that it must be explained to those who do not know it.

The best way to demonstrate the operation of this floral device is to lead one to a blooming milkweed where he may watch the bees, butterflies and other insects attempting to get nectar from the cups that make up what is known as the crown of the flower. There are five of these cups on each flower and they usually are found hanging mouth downward. They are very smooth and a foot of the insect that tries to alight on one of them invariably slips off and lands in the slit which lies between two of the cups. This is exactly what should happen; the very purpose for which the flower was made. In that slit lies the male principle of the flower in such a position that it may never come in contact with the female organ unless it is pulled out and pulled in again. So the foot of the insect goes into the slit and finds itself caught as if between two tough little wires. If the insect is strong enough a jerk or two will bring out the foot, and along with it two little bags of pollen which ride away on the insect's leg to another flower where the foot again slips into a slit, but this time carrying the pollen bags (pollinia, as this sort of pollen is called) right down upon the stigmatic surface of the female organ. There the bags stay and the pollen in them develops as it should, and in due time a whole podful of milkweed babies, so angelic that they can fly, sail off to do their weedy part in the plant kingdom.

Of course that same insect leg may come out of the second flower slit with two more bags of pollinia hanging to it like two little saddle bags. This may occur if another insect has not been there before him and had the same experience that he had when he visited the first flower. It takes a strong leg to do this work and strong insects are sometimes seen with several bags of pollinia hanging to their legs. If the insect is not strong enough he pays dearly for the sip of nectar he gets. The wages of sin is death in this case. No matter how worthy or how timidly beautiful the little insect is, and no matter if this attempt to steal a sip of nectar is his first offense, if he cannot pay for that nectar by carrying away to another flower those two little pollinia bags, he hangs there until he dies and returns to Mother Earth with the wilted and disappointed floral parts.

Just how the milkweed flower is constructed is succinctly told in the technical description. If the above seems too verbose and imaginative the reader will find the facts shorn of all romance in those brief lines at the end of this account.

This particular milkweed does not have a great many common names. It is called silkweed, cottonweed, silky swallow wort, Virginia silk, and wild cotton, the names all referring to the

FIG. 58. The Milkweed with flower enlarged
to show the insect trap

downy-winged seeds. The botanical name could be applied to any other weed. The generic part of it is given in honor of a Greek physician, Æsculapius. The specific name *syriaca* refers to Syria, perhaps because it was thought to have come from there.

The plant is a perennial and for that reason should not molest cultivated fields, but it does. Its stem penetrates the earth to a considerable depth and acts as an upright rootstalk. When the plant is pulled up the part of the rootstalk that remains in the ground sends up one or more shoots. The same thing happens when the weed is plowed out. The rootstalk is stored with food and bears buds. The weed has horizontal creeping rootstalks also, and these, often below plow depth, succeed in developing a patch of milkweeds from a single parent stalk right in the midst of a cultivated field.

The young stalks pull up readily and resemble blanched asparagus. It is said that they are cooked and eaten by some people. The plant at one time was in the United States Pharmacopœia and its extract asclepias is still obtainable and is still to be found listed in the United States Dispensatory. This extract is used for asthma, dyspepsia, and coughs.

TECHNICAL DESCRIPTION

Asclepias (Tourn.) L. Calyx persistent divisions small, reflexed. Corolla deeply 5-parted; divisions valvate in bud, deciduous. Crown of 5-hooded bodies seated on the tube of stamens, each containing an incurved horn. Stamens 5, inserted on the base of the corolla; filaments united into a tube which encloses the pistil; anthers adherent to the stigma, each with 2 vertical cells, tipped with a membranaceous appendage, each cell containing a flattened pear-shaped and waxy pollen-mass; the two contiguous pollen-masses of adjacent anthers, forming pairs which hang by a slender prolongation of their summits from 5 cloven glands that grow on the angles of the stigma (extricated from the cells by insects, and directing copious pollen-tubes into the point where the stigma joins the apex of the style). Ovaries 2, tapering into very short styles; the large depressed 5-angled fleshy stigmatic disc common to the two. Follicles 2, one of them often abortive, soft, ovoid or lanceolate. Seeds anatropous, flat, margined, bearing a tuft of long silky hairs (*coma*) at the hilum, downwardly imbricated all over the large placenta, which separates from the suture at maturity. Embryo large, with broad foliaceous cotyledons in thin albumen.—Perennial herbs; peduncles terminal or lateral and between the usually opposite petioles, bearing simple many-flowered umbels, in summer. (The Greek name of *Æsculapius*, to whom the genus is dedicated.) *Asclepias syriaca* L. Stem tall and stout, finely soft-pubescent; *leaves* lance-oblong to broadly oval, 1-2 dm. long, pale, *minutely downy beneath, as well as the peduncles,* etc.; corolla-lobes dull purple to white, 6-9 mm. long; *hoods* rather longer than the anthers, *ovate, obtuse, with a tooth each side of the short stout clawlike horn.* Rich ground, New Brunswick to Saskatchewan, and southward. June–August.— Intermediates, perhaps of hybrid origin, occur between this and some of the related species.

VINING MILKWEED

[*Gonolobus lævis* Michx.]

ONLY the botanist would ever suspect that the Vining milkweed is a milkweed. It has the milkweed flower and pod, but that is all. There is no milk and it behaves like a pole bean. Perhaps that is why it is called pea vine in some localities. In the popular mind there seems to be little distinction made between peas and beans and so any vine that resembles either the pea or bean is likely to be called a pea vine.

Because the weed is a perennial with a rootstalk placed far below plow depth, it is able to thrive in cultivated fields, where it often becomes as much of a pest as the morning glories and bindweeds. The vine unmolested will get to be from fifteen to twenty feet long, and although it does not bear many pods a single pod will scatter to the winds hundreds of seeds. It is the milkweed way that the vine has, and nothing but a dandelion can equal the distribution method of the milkweed. It depends upon the wind, but a current generated by a passing bee's wing is enough to set the seeds afloat.

This effective seed distribution method coupled with a well-placed perennial rootstalk gives to the vining milkweed an advantage seldom met with even in the worst of weeds. It is on a par in these respects with the Canada thistle, and it is not so climate-bound as is that pest of the northern States. It is fast occupying regions where it was unknown a few years ago. It is so new in some localities that it is confused with bindweeds, with the white swallow-wort, and with a weed never seen except in waste places: the wild buckwheat vine. There is a slight resemblance in all mistaken identities, but none of the vines with which the vining milkweed is confused has its deep green, long, heart-shaped leaves, and the little elongated clusters of flowers in the leaf axes separate it from all the other vines except, perhaps, the white swallow-wort, though the white swallow-wort's leaves are not heart-shaped.

The common names of this weed are Vining or Climbing milkweed, Sand vine, and Angle-pod. The generic name, *Gonolobus*, is made up of two Greek words meaning angle and pod. The specific name, *lævis*, probably means winding to the left, referring to the way the vine climbs.

TECHNICAL DESCRIPTION

Gonolobus Michx. Crown of free leaflets, which are truncate or obscurely lobed at the apex, where they bear a pair of flexuous awns united at base. Anthers nearly as in *Asclepias;* pollen-masses oblong, obtuse at both ends, fixed below the summit of the stigma to the descending glands. Follicles elongate-ovoid to lanceolate, smooth. Seeds with a tuft, as in *Asclepias.*—A perennial twining herb, smooth, with opposite heart-ovate and pointed long-petioled leaves, and small whitish flowers in racemelike clusters on slender axillary peduncles. (Name from two Greek words meaning *an angle* and *a pod,* from the angled fruit.)

Gonolobus lævis Michx. Climbing, 3–4 m. high; leaves 3.5–12 cm. wide.—River-banks and thickets, Pennsylvania to Illinois, Kansas, and southward. July–September.

FIG. 59. The Vining Milkweed is milkless

THE MORNING GLORY

[*Ipomœa purpurea* L.]

IN SPITE of the beautiful flowers that truly glorify a summer's morning the common Morning glory is a hateful weed in many localities. It likes good soil and when it is once established in a rich bottom field or in the rich brown silt loam of the prairies it is a nuisance, to say the least. The vines wind themselves around the growing cornstalks, soy beans, cotton plants, and even around weeds that should be able to cope with such weak little plants as the Morning glories seem to be. But they are not weak. Like snakes, those slender vines crawl up over the plants they select for their trellises, and soon the big Morning-glory leaves are shading the leaves of the trellising plants, and very soon after that those glorious flowers will be smiling on all the world like a big woman obstructing the view of a small boy at the movies.

There are three annual Morning-glory species that infest fields and gardens throughout the greater part of the United States, and two of these are so nearly alike that no one but the botanist needs consider their differences. One main difference in these two can be seen at a glance, however. The leaves of the one treated in this sketch are heart-shaped and entire; the leaves of the other one are three-lobed. The leaves of both of these species are represented in the drawing.

The botanical name of the three-lobed species is *Ipomœa hederacea* Jacq.; the botanical name of the third species is *Ipomœa coccinea* L., and it can be distinguished from the first two by its smaller and light scarlet flowers. The flowers of the other two species are purple, blue, and white.

The name, *Ipomœa,* is from the Greek and means wormlike, and it takes no stretch of imagination to see the wormlikeness in the crawling vines of the Morning glory. *Purpurea,* the specific name of the species here treated, means purple. So *Ipomœa purpurea* L. means the purple-flowered plant that crawls like a worm, named by Linnæus.

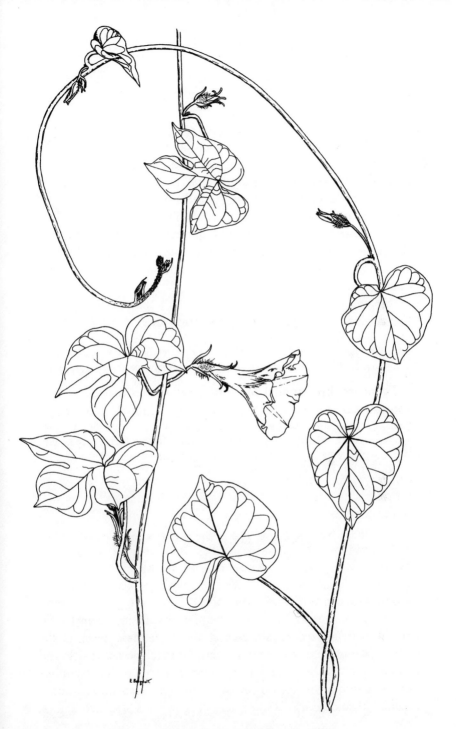

FIG. 60. Two species of the Morning Glory

As has already been stated, the common Morning glory flourishes only in good soil. It is a fine indicator of sweet and sour soil, for the little plants are often dwarfed and made yellow by sour soil. They are also often dwarfed and made yellow by a rust disease. The rust pustules break out on the lower side of the leaves, and if one finds yellow, dwarfed Morning-glory plants in his field or garden he should look at the underside of the leaves before he decides that his ground is sour.

Like all fast-growing, succulent weeds the young Morning-glory vines make good fertilizer, and the gardener or farmer can well smile in his turn as he sees his hoe or plow pull these glorious vines into his soil, where they will become the food of the soil-building bacteria.

TECHNICAL DESCRIPTION

Ipomœa L. Calyx not bracteate at base, but the outer sepals commonly larger. Corolla salver-form or funnel-form to nearly campanulate; the limb entire or slightly lobed. Capsule globular, 4–6 (by abortion fewer)–seeded, 2–4–valved.

Ipomœa purpurea (L.) Roth. Annual; stems retrorsely hairy; *leaves heart-shaped, acuminate, entire;* peduncles long, umbellately 3–5–flowered; calyx bristly-hairy below; corolla funnel-form, 4.5–7 cm. long, purple, varying to white.—Escaped in cultivated grounds. (Introduced from Tropical America.)

MAN-UNDER-GROUND

[*Ipomœa pandurata* G. F. W. Mey.]

"MAN-UNDER-GROUND" and "Man-of-the-earth" are strange names for a plant, but no stranger than the plant itself, since it insists on placing its man-sized storage root at about the same depth in the earth as we place our lamented dead. Storage roots as thick and as long as a man's leg and weighing from ten to thirty pounds may be found at a depth of from six to eight feet under a long-established colony of this most persistent weed. These two names,

FIG. 61. The above-ground expression of the Man-under-ground

"Man-under-ground" and "Man-of-the-earth," are descriptive enough to him who has seen the enormous storage roots, but some of the other names are more helpful to those who have seen only the ærial expression of this unique phytological personality. In other words, the vines that grow like sweet potatoes and bloom like morning glories are all that 90 per cent of those who see the weed ever observe of it, and this includes botanists as well as laymen. So the names Potato vine, Wild potato vine, Wild sweet potato vine, Wild morning glory, and Bindweed are all better names for the plant than the first two given. It is also called Mecha-meck (an Indian name) and Wild jalap. It is said that an inferior jalap is extracted from its storage root.

There are several morning glories and bindweeds; so this particular one needs to be described. It is the one usually found in fields and waste places where the soil is deep and inclined to be wet and heavy. It is usually found hobnobbing with the shoestring smartweed since these two weeds like exactly the same kind of soil: deep, rich, but often wet. One may find other bindweeds in such soil, but this is the one with the heavy, reddish-tinged stems and with smooth, heart-shaped, sometimes fiddle-shaped leaves. The growing ends of the vines are filled with a sticky, milky juice that should have set the rubber chemists to work on them long before this. This plant may be a valuable source of rubber.

Where there is nothing for the vines to climb they spend their time in covering the ground and in spreading their white, funnel-shaped flowers to the morning sun and to the moths and bumble bees. The throat of the flower has a pinkish-purplish color.

In some places the foreign laboring class has learned to make hay of the weed by hanging it on fences to dry. Their cows are very fond of this hay and are said to thrive on it.

Man-under-ground is one of the most persistent weeds. It never quits sending up vines until the "man" is exhausted, and a full grown "man" can hold out for three years. Where a patch of the weed has become established one can completely eradicate it by visiting the patch every week throughout one growing season, pulling out the young vines on every visit. The next year will not require so many visitations, but the vines sent up must not be permitted to set many leaves. There will be not nearly so many vines to pull the second year as the first, and yet there will be far too many to ignore. If the work has been well done the first and second years, a single pulling on the third year will likely end the patch. If no more vines appear, the men-under-ground have been starved to death, and the work of eradication is complete.

According to Linnæus the name *Ipomœa* is from two Greek words which when put together mean "like a bindweed." *Pandurata* means shaped like a pandore, an instrument that was fashioned something like a violin. Some of the leaves of the vine have this fiddle shape, and so the name, *pandurata*.

TECHNICAL DESCRIPTION

See Morning Glory for technical description of *Ipomœa* L.

Ipomœa pandurata (L.) G. F. W. Mey. Perennial, smooth or nearly so when old, trailing or sometimes twining; leaves occasionally contracted at the sides so as to be fiddle-shaped; *peduncles longer than the petioles,* 1–5-flowered; *sepals smooth, ovate-oblong, very obtuse;* corolla open-funnel-form, 4.5–8 cm. long, white, with purple in the tube.—Dry ground, Connecticut to Ontario, southward and southwestward. June–September.—Stems long and stout, from a huge root, which often weighs 4–8(–11) kg.

WILD MORNING GLORY

[*Convolvulus sepium* L.]

THE best common name this plant has is Hedge bindweed. It is usually seen in hedges and along fence rows, and it is a bindweed, not a morning glory. What the difference between morning glories and bindweeds is makes little difference to the layman, but this is a bindweed, a *Convolvulus,* while the morning glory is an *Ipomœa*. The weed can easily be identified by its hastate (arrow-head-shaped) leaves and by its pale pink, striped with white, morning-glory-shaped flowers.

The wild morning glory is a true weed but not the equal of its cousin, the Creeping Jenny, when it comes to weediness. It enters hayfields and meadows, but it is easily eradicated by plowing. The Creeping Jenny is only scattered and set by plowing. The Wild morning glory is at its best in unmolested hedge and fence rows, and one does not have to go farther than the nearest overgrown fence row, especially where he knows the soil along that fence row is inclined to be wet, to find this species of bindweed. It may become a real pest in old hayfields and meadows, and it stays on for a year or two in cultivated fields and gardens when those fields and gardens have been taken from infested meadows and lots. It does not stay long in well-cultivated soil, however. It has rootstalks just as the Creeping Jenny has, but evidently they are unable to withstand the torture that those of the outlaw cousin are able to endure.

The plant has a great many common names. It is called Hedge bindweed, Great bindweed, Bell bind, Woodbind, Pear vine, Devil's vine, Lady's nightcap, Hedge lily, Harvest lily, Rutland beauty, Creepers, Bracted bindweed, Woodbine, and German scammony.

The botanical name, *Convolvulus sepium,* means Hedge bindweed. *Convolvulus* is from the Latin, *convolvere,* meaning to entwine, and *sepium* is from the Latin *sæpes,* meaning a hedge. This weed is a sepicolous plant.

TECHNICAL DESCRIPTION

Convolvulus (Tourn.) L. Corolla funnel-form to campanulate. Stamens included. Capsule globose, 2–celled, or imperfectly 4–celled by spurious partitions between the 2 seeds, or by abortion 1–celled, mostly 2-4-valved.—Herbs or somewhat shrubby plants, twining, erect, or prostrate. (Name from *convolvere,* to entwine.)

Convolvulus sepium L. *Glabrous or essentially so;* stem *high twining or sometimes trailing extensively; leaves triangular-halberd-shaped,* acute or pointed, the basal lobes obliquely truncate and often somewhat toothed or sinuate-lobed or merely rounded; *peduncles* chiefly *elongated,* 4–angled; bracts rounded to sharp-acuminate at tip; corolla white or rose-color, 3–5 cm. long.—Moist alluvial soil or along streams. June–September. (Eurasia.)

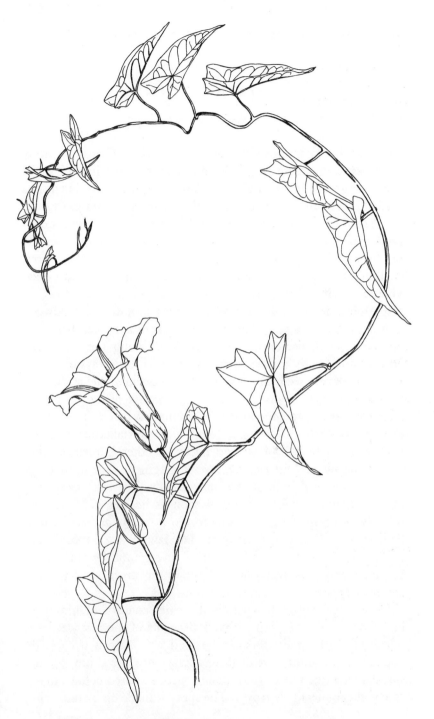

Fig. 62. The Wild Morning Glory or the Hedge Bindweed

CREEPING JENNY

[*Convolvulus arvensis* L.]

CREEPING JENNY is one of the meanest of weeds. That name aptly describes it. A whispering little hussy that creeps in and spoils everything. The weed needs no other name than this, but it has several others. Its flower resembles that of the morning glory, and so we could expect it to be called Wild morning glory, and Small-flowered morning glory. It is also called Bindweed, European bindweed, Field bindweed, Hedge bells, Corn-lily, Withwind, Bellbine, Lap-love, Sheep-bine, Corn-bind, Bear-bind, and Green vine.

Creeping Jenny is a bindweed; the small-flowered bindweed that does more creeping than binding. It is usually seen lying flat on the ground, flowering and seeding all along the line. Like all perennials with horizontal rootstalks it has two ways of propagating itself: by seeds and by rootstalks. Luckily the seeds have no special device for dispersal, but the rootstalk method is highly effective, especially in cultivated ground. When a plow tears through a patch of Creeping Jenny the rootstalks are torn up and scattered by the plow, and by every other implement that follow the plow. And every rootstalk fragment that finds lodgment in the earth will start a new colony. For this reason the farmer should know the weed as soon as he sees it. If he does not have it on his farm he is due for an invasion, and like Uncle Sam he should be prepared for it. The lawn owner, too, should make the acquaintance of Creeping Jenny before she appears on his lawn, and so be ready to land her in the street or alley when she does appear.

How to get rid of the pest is the chief concern of him who has to deal with Creeping Jenny. This is one of the few weeds on which chemical eradicators can be used with success if correctly applied. It is said that two or three applications of sodium chlorate sprayed on at the rate of from two to three hundred pounds to the acre will completely destroy it. The spray is made up by dissolving

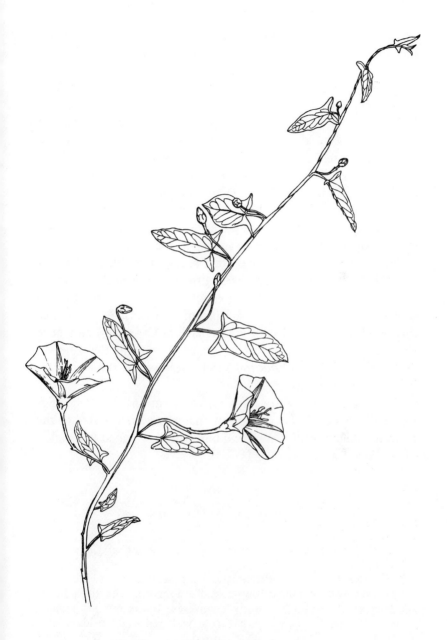

Fig. 63. Creeping Jenny is a federal outlaw

a pound of the chemical in each gallon of water used. The second and third sprays need not be so heavy as the first, and it is well to wait until the second year to apply the third spray. But why use a spray unless the weed has taken possession of a considerable area? Small patches can and should be eradicated with a hoe or by hand pulling. If the patch is in a lawn where the hoe cannot be used without doing considerable damage to the grass the chlorate may be resorted to if care is taken in applying it. It must be poured on the weed only. Chlorate kills grass as well as weeds. Hand pulling is laborious, but it is just as effective, if followed up, as is the chemical method. Mechanical methods depend upon the fact that the leaves of the plants make the food that is stored in the rootstalks, and upon the additional fact that newly developed shoots draw on that food-storage supply. Hoeing off or pulling up those shoots every time they put in their appearance will finally exhaust the fattest of roots and rootstalks. Persistence is all it takes to destroy perennial weeds. A persistence that will make the eradicator visit a patch of Creeping Jenny once each week for a single growing season, removing the vines in sight on visitation, will destroy, in one year's time, the best-established patch of the weed.

Creeping Jenny is a *Convolvulus,* which means that it belongs to the genus of twiners. Its specific name, *arvensis,* refers to the fields in which it grows, so "Field bindweed" is a good translation of its botanical name.

TECHNICAL DESCRIPTION

See Wild Morning Glory for technical description of *Convolvulus* (Tourn.) L.

Convolvulus arvensis L. Perennial; stem procumbent or twining, and low; leaves ovate-oblong, arrow- or halberd-shaped, with the lobes at the base acute; peduncles mostly 1-flowered; bracts minute, remote; corolla 1.5–2 cm. long, white or tinged with red.—Old fields and in waste places. June–August. (Naturalized from Europe.) Var. *obtusifolius* Choisy. Basal lobes of the leaves rounded.—Less common. (Adventive from Europe.)

WHITE VERVAIN

[*Verbena urticæfolia* L.]

WHITE VERVAIN or Nettle-leafed vervain is one of a whole genus of weeds. It is true that two species of the genus are progenitors of the verbenas of the flower garden, but they, too, are weeds, and the rest of the species are all worthless, homely, stout-stemmed plants. All but one of the species are perennials and prone to establish themselves in overcropped pastures and waste places. The White vervain is perhaps the most worthless and homeliest of the lot.

The name, vervain, seems to be a corruption of verbena. Verbena is from the Latin and according to *Webster's New International Dictionary,* Second Edition, Unabridged, the derivation of the word is very different from that found in the herbals, which is given in this book under the caption, Blue Vervain. Here let it be said that according to the dictionary the word is probably derived from the Latin word *verber,* which means rod, stick, or stem. When one knows this fact and also knows the vervains he cannot keep from admiring the erudition of Linnæus, who gave the name *Verbena* to the genus. Most of the species are all stems. It is because of the tough, stemmy growth of the white and blue vervains that country boys make brushes of the weeds when they go out to fight bumblebees. The same stemmy character adapts these weeds to the sprinkling of holy water. The vervains have been used as hyssop for hundreds if not for thousands of years.

The White vervain is too woody to make a good fertilizer and too small to serve as wood. The stems are tough enough and slender enough, however, to be used as arrows, and they have been so used by many a boy, both red and white. But it was not to this end that these slender, woody stems were made. The hundreds of seeds which develop at the top of them have no means of dispersal other than the wind's swaying and pitching of these strong,

Fig. 64. The White Vervain

lithe stems. Of course some of the seeds are eaten by birds, and so some service is rendered by the White vervain, but the quantity of service is nothing in comparison with the valuable space the weed occupies throughout the entire year. The dead stems of the year before are usually standing when the worthless growth starts in the spring.

TECHNICAL DESCRIPTION

Verbena (Tourn.) L. Calyx 5-toothed, one of the teeth often shorter than the others. Corolla tubular, often curved, salver-form; the border somewhat unequally 5-cleft. Stamens included; the upper pair occasionally without anthers. Style slender; stigma mostly 2-lobed.—Flowers sessile, in single or often panicled spikes, bracted, produced all summer. (The Latin name for any sacred herb; derivation obscure.)— The species present numerous spontaneous hybrids.

Verbena urticæfolia L. Perennial, from minutely pubescent to almost glabrous, rather tall (0.5–1.5 m. high); *leaves oval or oblong-ovate, acute, coarsely serrate, petioled;* spikes at length much elongated, loosely panicled; *flowers* very small, *white.*—Thickets, roadsides, and waste ground. (Tropical America.)

BLUE VERVAIN

[Verbena stricta Vent.]*

THERE ARE two blue vervains and this is not the one the botanist has in mind when he says "blue vervain." That plant is the tall one found in damp places; this one is the short, robust weed that grows in hot, dry places as well as in the habitat of the other species. This one is often called Hoary vervain, Mullein-leafed vervain, and Woolly vervain, but its flowers are blue and so it deserves to be called Blue vervain.

It is more of a weed than its cousin, *hastata,* the true blue vervain, for it is able to negotiate a wider range of habitat than its cousin. It is at its best in overgrazed pastures and feed lots where it hobnobs with Yarrow and Ironweeds (Vernonias) and with another cousin, the White vervain.

Blue vervain is easily identified. When one sees a harsh herb with a square, woody stem crowned with several straight, blunt spikes, around each of which is a circle of small blue flowers that climbs the spike as the days progress, he is seeing the Blue vervain. The country boy makes "swatters" of the weeds when he goes out to fight bumblebees, and the dead stems make arrows for his hastily constructed bows. In other words the woody nature of the plant makes "ironweed" rather descriptive, and its bushy character gives to it the name of Wild hyssop. It is also called Purvain.

According to *A Modern Herbal*, the name vervain is from the Celtic "ferfaen"—*fer* meaning to drive away, and *faen,* a stone. It was thought to be effective in cases of gravel (Calculus), and in bladder and urinary troubles. The name verbena, the botanical name, was used by the Romans to designate altar plants. The specific name, *stricta,* refers to the straight (strict) stems and flowering spikes.

The bruised leaves of the weed are said to relieve headache, earache, and rheumatism. The tea is used externally for piles and internally as a purgative. How such weeds ever came to be re-garded medicinally is a mystery to him who knows the plant. It does not have a single character that would lead a sick person to discover it. Perhaps the fact that the vervains were used as altar plants—as hyssop to sprinkle holy water—led the common folk to test their cleansing virtues. Surely these weeds would drive out the devil actually if they were used figuratively for that purpose. At any rate vervain tea was drunk to expel gravel, and its bruised leaves were placed on aching heads and ears and around rheumatic joints.

Blue vervain is native to this country and is so very common in the Mississippi Valley that a midsummer rural landscape in that section without its vervains and ironweeds would hardly be complete. The plant is a perennial and so does not molest cul-tivated fields to any great extent, but in pasture land, where the

FIG. 65. A Blue Vervain, but not the
Blue Vervain of botanists

grazing stock never touch it, and along highways, it thrives even if the location is rather dry.

TECHNICAL DESCRIPTION

See White vervain for technical description of *Verbena* (Tourn.) L. *Verbena stricta* Vent. *Downy with soft whitish hairs,* erect, simple or branched, 3–9 dm. high; *leaves sessile, obovate or oblong, serrate; spikes thick,* somewhat clustered, hairy; flowers rather large, purple.— Barrens and prairies, Ontario and Ohio, westward and southward; rarely native eastward.

HOREHOUND

[*Marrubium vulgare* L.]

If one wants to see horehound he must go where it grows, in waste places, neglected and unused lots, old worn-out pastures, and even in sparsely wooded areas where its relative, the Pennyroyal, holds sway. The weed is a perennial and yet is never obtrusive. It never becomes a weed in cultivated fields, and is easily removed even from its most successfully established stronghold. Remove a bunch of horehound, roots and all—and that is easily done—and another more aggressive weed will take its place. The plant has little fight in it even though it is beautifully equipped to scatter its seeds and to ward off its enemies. The seeds grow readily enough, too, but the seedlings must be unmolested if they are to become adult plants.

Then why treat this plant as a weed? Because of its taste and smell. If horehound once gets into a pasture it can stay there, since grazing animals never touch it. It is bitter, and although its odor is not offensive to man it seems to be so to stock. Horehound is one of the medicinal plants, and like all plants that have made Pharmacopœia it was first selected because of its bitter taste and rank odor.

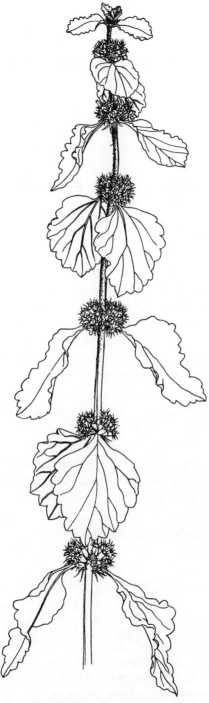

Fɪɢ. 66. The Horehound

The weed actually has some medicinal value. Its tea or extract is still used to flavor and color candy, which confection has a soothing effect on irritated throats. Strong horehound tea acts as a purgative (a thing that is always effective in combating colds) and it is also used as a vermifuge.

The botanical name of the plant, *Marrubium,* is taken from the Hebrew and means bitter; *vulgare,* the specific name, means common. It is the common bitter plant, according to its botanical name. Like most plants it has several English names: Hounds-bene, Marrube, and Marvel are three of them. But horehound is the name used most and so deserves to be defined. It means the gray plant, and not the gray dog as one may be led to think. The *hore* is the same as *hoar,* which means gray. The *hound* part of the name has a strange derivation, too long to be given here, but it finally reaches the point where it means plant. Horehound is a good name for the plant since its stems are gray or hoary, made so by the hairs. In the technical description we find "Whitish-woolly" given to describe its appearance.

Horehound has two peculiar characters that deserve special mention: it is one of the few mints that is evergreen, and it has a mode of seed distribution that is very unusual. The calyx of the flower becomes a little seed basket that is rimmed with very effective little hooks for attaching the basket to hair and clothing. If the reader has never seen the plant he should look for it in the winter time, for then as sure as he finds a deep-green, strong-smelling, bitter tasting mint with white hairs on its stems and leaf petioles, and with little seed baskets that cling to his gloves or mittens, he finds the horehound.

TECHNICAL DESCRIPTION

Marrubium (Tourn.) L. Calyx-teeth more or less spiny-pointed and spreading at maturity. Upper lip of the corolla erect, notched, the lower spreading, 3–cleft, its middle lobe broadest. Stamens 4–Whitish-woolly bitter-aromatic perennials, branched at the base, with rugose and crenate or cut-leaves, and many-flowered axillary whorls. (A name used by Pliny, from the Hebrew *marrob,* a bitter juice.)

Marrubium vulgare L. Stems ascending; leaves round-ovate, petioled, crenate-toothed; whorls capitate; calyx with 10 recurved teeth, the alternate ones shorter; corolla small, white.—Waste places, Maine to Ontario, westward and southward. June–August. (Naturalized from Europe.)

CATNIP

[Nepeta cataria L.]

CATNIP TEA is a diaphoretic and an emmenagogue. The catnip itself is a very common weed in spite of the fact that the dictionary has to be resorted to to learn the use of its tea. The plant grows in waste places, old garden patches, hedgerows, vacant lots, anywhere soil is good, throughout the northern part of the United States and the southern part of Canada. It is found in some of the southern States, also, but wherever it is found it seldom becomes a bad weed.

The catnip is a tall, straight-stemmed mint, reaching a height of four feet in rich soil, and it is the one tall mint that is downy. The square stems and the soft, downy leaves that are heart-shaped with scalloped edges are enough to identify the plant, but it has another characteristic that will identify it even in the dark. That is the odor of the crushed leaves. After one has smelled that odor his nose knows catnip. And it is the odor of the crushed or bruised leaves that attracts the cats and so gives the weed its name. Cats eat it after they have been attracted to it. In Europe, where catnip is raised for the market, they have a rhyme that reads:

"If you set it the cats will get it,
If you sow it the cats won't know it."

This is true because the handling of the plants in setting them bruises them enough for the cats to detect the odor. When the plants come up as seedlings the cats never discover them.

Because catnip is a medicinal plant (it was once placed in the Pharmacopœia) it is well to know that home remedies can be and

FIG. 67. The Catnip

are made from it. Tea from either the fresh or dried leaves will produce perspiration and for this reason it is good for colds. The tea will induce sleep in fever patients, and it has been used for years in cases of scarlet fever and smallpox. It will relieve colicky pains, and is enough of a sedative to be used in cases of hysteria.

Catnip is one plant that has few common names. It is sometimes called Cat mint, and the English also call it Catnep. The meaning of the botanical name, *Nepeta cataria* L., is the Latin way of saying "The cat plant from Nepete," Nepete being an Etruscan city in northern Italy in the days of the Roman Empire.

TECHNICAL DESCRIPTION

Nepeta L. Calyx tubular, often incurved. Corolla dilated in the throat; upper lip erect, rather concave, notched or 2–cleft; the lower 3–cleft, the middle lobe largest, either 2–lobed or entire.—Perennial herbs. (The Latin name, thought to be derived from Nepete, an Etruscan city.)

Nepeta cataria L. Downy, erect, branched; leaves heart-shaped, oblong, deeply crenate, whitish-downy underneath; corolla whitish, dotted with purple.—Near dwellings; a common weed. July–September. (Naturalized from Europe.)

GILL-OVER-THE-GROUND

[*Nepeta hederacea* L.]

AN INTERESTING little weed that is sometimes a very bad little weed is Gill-over-the-ground. It is one of the weed emigrants and it brings with it several names from the days of Merrie England. There it was called Gill-over-the-ground, Haymaids, Tunhoff, Hedgemaids, Lizzy-run-up-the-hedge, Cat's-foot, Gill-go-by-the-hedge, Creeping-Charlie, Robin-run-in-the-hedge, and Ground ivy.

It is a mint but without the pronounced minty odor and taste that most known mints have. The square stems, the lipped flowers, and the faint but true mint tang show that it is a mint, and keep it from being confused with any of the round-leafed, creeping plants that are not mints. The plant is related to catnip. Catnip and Gill-over-the-ground are cousins. They are in the same genus; they are both Nepetas, and so we might expect Gill-over-the-ground to be classed as medicinal. The bruised leaves do act as an astringent, and the tea is used as a tonic and as a gentle stimulant. It is said to be useful in kidney diseases and to help in cases of indigestion. Like all aggressive or in any way spectacular plants, more has been claimed for it by the herbalists than the weed was ever able to do.

There is only one other mint that is likely to be confused with Gill-over-the-ground, and that is the annual known as Dead nettle or Henbit. It is not so ivylike as Gill-over-the-ground, and yet its stems are decumbent and when it gets into gardens it does the same tricks there that Gill-over-the-ground does in lawns: it takes the place. The flowers of the two weeds are of about the same size and color and are born in the axes of the upper leaves, which nearly or quite surround the stems in both plants. But the flowers are not alike in shape. The flowers of the Dead nettle are more open, there is more of a throat and more of a mouth than in the flowers of Gill-over-the-ground, and although the Dead nettle's many branches are decumbent and may take root where they touch the ground the plant does not crawl along the ground as the other one does, and it starts from seed each spring, or fall if it is in the southern States. Dead nettle is much more active in the South than is Gill-over-the-ground.

The botanical name of Gill-over-the-ground, *Nepeta hederacea,* means the ivy from Nepete, but, of course, the generic name is given to this plant not because it came from Nepete, but because it belongs to the same genus that Catnip belongs to. *Nepeta cataria* probably was named in honor of the Etruscan city, Nepete, and since Gill-over-the-ground has the same generic characters that Catnip has (see technical description of Nepeta at the end of the write-up on catnip) it belongs to Nepeta, the catnip genus. The specific name, *hederacea,* means ivylike.

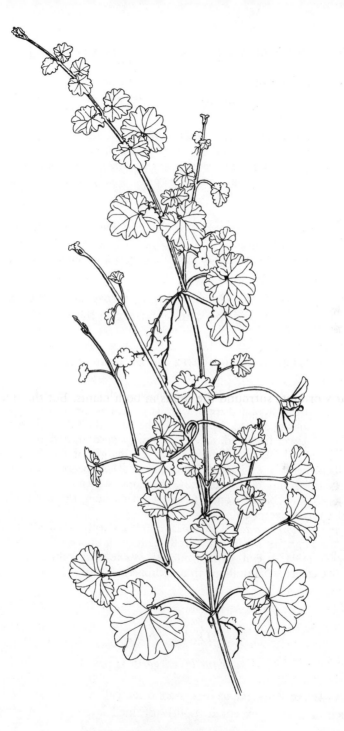

FIG. 68. Gill-over-the-ground as
it appears from above

TECHNICAL DESCRIPTION

See Catnip for technical description of *Nepeta* L.
Nepeta hederacea (L.) Trevisan. Creeping and trailing; leaves petioled, round-kidney-shaped, crenate, green both sides; corolla thrice the length of the calyx, light blue.—Damp or shady places, near towns. May–July. (Naturalized from Europe.)

Since it has been necessary to describe Dead nettle or Henbit, the technical description should be given here also. The botanical name is *Lamium amplexicaule* L., which, freely translated, means the open-throated flower with leaves that surround the stem. *Lamium* is from the Greek, meaning throat, and *amplexicaule* means clasped around the stem—referring to the leaves.

TECHNICAL DESCRIPTION OF DEAD NETTLE

Lamium L. Calyx tubular-bell-shaped, about 5-nerved, with 5 nearly equal awl-pointed teeth. Corolla dilated at the throat; upper lip ovate or oblong, arched, narrowed at the base; the middle lobe of the spreading lower lip broad, notched at the apex, contracted as if stalked at the base; the lateral ones small, at the margin of the throat.—Decumbent herbs, the lowest leaves small and long-petioled, the middle heart-shaped and doubly toothed, the floral subtending the whorled flower-cluster. (Name from a Greek word meaning *throat,* in allusion to the ringent corolla.)
Lamium amplexicaule L. Leaves rounded, deeply crenate-toothed or cut, the *upper ones clasping;* corolla elongated, upper lip bearded, the lower spotted, lateral lobes truncate.—Waste and cultivated places. April–October. (Naturalized from Europe.)

HEAL-ALL

[*Prunella vulgaris* L.]

HEAL-ALL or Prunella, as it is more often called, never amounts to much as a weed except in poorly drained lawns. It is a strange

FIG. 69. Heal-all, the plant that lives
all over the world

plant in that one seldom finds it in abundance anywhere and yet it is found everywhere. A man who had spent seven years in Japan and who had lived on nearly every island in the archipelago said that he never failed to find this little plant wherever he went, and that although he had been in many nations of the world he had never entered one where Prunella was not present to greet him—not many plants, but always enough to attract his attention. And so it is in America. *Gray's Manual of Botany* says: "Woods and fields, Newfoundland to Florida and westward across the continent."

There seems to be no reason why the plant was called Heal-all and Self-heal. It never was used as a cure-all. And why is it called Carpenter weed? There is a weed called Carpenter's square which has a big, square stem with leaves extending out at right angles to it, but there is nothing about this little weed that would suggest a use in the carpenter's trade. The only name it has that comes near telling the truth about the plant is its generic name. *Prunella* is evidently the way Linnæus heard its German name *Brunella,* a word which was intended to indicate that the weed was used in cases of *Bräune,* the German name for quinsy. That was one ailment the plant was actually used for. Its leaves were bruised and bound on throats suffering with quinsy. The *U. S. Dispensatory* says, "It was formerly used in hemorrhages and diarrhœa and as a gargle in sore throat." The hemorrhage and diarrhœa use was derived from the Indians. They used it "in cases of dysentery, especially for babies."

As a lawn weed, and especially in poorly drained lawns, it may become very troublesome. But even if the reader never sees it in his lawn he ought to know the plant. He ought to know that here is a plant that grows in every part of the world. He ought to know that this is a mint, but one with every minty character so modified that he may be forgiven for not recognizing its family traits. He ought to know that here is a modest little thing with almost no medicinal virtues that is called Heal-all and Self-heal, for in knowing that, he may be able to sense how far civilization has come in at least one phase of its development.

TECHNICAL DESCRIPTION

Prunella L. Calyx tubular-bell-shaped, somewhat 10–nerved, naked in the throat, closed in fruit; upper lip broad, truncate. Corolla ascending, slightly contracted at the throat and dilated at the lower side just beneath it, 2–lipped; upper lip erect, arched, entire; the lower reflexed-spreading, 3–cleft, its lateral lobes oblong, the middle one rounded, concave, denticulate. Filaments 2–toothed at the apex, the lower tooth bearing the anther; anthers approximate in pairs, their cells diverging. —Low perennials, with nearly simple stems, and 3–flowered clusters of flowers sessile in the axils of round and bractlike membranaceous floral leaves, imbricated in a close spike or head. (Name said to be from the German *Bräune,* a disease of the throat, for which this plant was a reputed remedy. Often written *Brunella,* which was a pre-Linnean form.)

Prunella vulgaris L. Leaves ovate-oblong, entire or toothed, petioled, hairy or smoothish; corolla violet or flesh-color, rarely white, not twice the length of the purplish calyx.—Woods and fields, Newfoundland to Florida, westward across the continent. June–September. (Europe.) Var. *laciniata* L. Some of the upper leaves tending to be pinnatifid.

MOTHERWORT

[*Leonurus cardiaca* L.]

"There's sumpin' you art to know," said an old midwife to a young bride as the two of them stood looking at a tall mint growing by the garden fence. "That's Motherwort."

One of the tallest of the mints is the Motherwort. It is likely to be found near dwellings either in the country or the town. It is a perennial, and like most perennials it cannot thrive in cultivated ground, but if left unmolested in good soil it will grow to as much as six feet in height. A tall, sparsely branched mint with deeply cut, palmate leaves (the lower ones appearing very much like small maple leaves), and with clusters of either flowers or seed baskets at the base of each pair of opposite floral leaves on the long spikelike tops—that is the Motherwort.

Wort is Anglo-Saxon for plant, so Mother wort is simply mother plant. That is what it is and what it has been for hundreds of years. Its leaves and flowers are used as an emmenagogue, a diaphoretic, a tonic, and a nervine, and so they furnish just what is needed by her whose nerves become frayed in certain stages of pregnancy. And even before pregnancy is known, a few draughts of Motherwort tea are said to dispense with all uncertainty in the matter by bringing to pass, or by failing to do so, the sign that every woman knows.

It is the tea made from the leaves and flowering shoots, either fresh or dried, that has all of the medicinal virtues mentioned above. The plant seems to have nothing that the Catnip does not have, however, and since catnip is of use to cats as well as to the female of our species, catnip is the popular weed. It is the one that is grown for the market.

Motherwort has few other common names. It is called Lion's-tail, Lion's-ear, and Cowthwort in England. The botanical name, *Leonurus,* means lion's-tail, and *cardiaca* probably refers to the heart-shaped leaves. The lower leaves would be heart-shaped if they were entire. A line drawn around the leaf just touching the tips of the lobes will make a heart-shaped outline. The fact that the leaves are lobed sets this plant apart from all of the other mints, except the two other species in the same genus. Motherwort is the attractive one of the genus, and the one that has personality enough to find a place in this book.

TECHNICAL DESCRIPTION

Leonurus L. Calyx 5-nerved, with 5 nearly equal teeth. Upper lip of the corolla oblong and entire, somewhat arched; the lower spreading, 3-lobed, its middle lobe larger, narrowly oblong-ovate, entire, the lateral ones oblong.—Upright herbs, with cut-lobed leaves, and close whorls of flowers in their axils; in summer. (Name made up of two words from the Greek meaning *lion* and *tail.*)

Leonurus cardiaca L. Tall perennial; *leaves long-petioled, the lower rounded, palmately lobed, the floral wedge-shaped at base, subentire or 3-cleft,* the lobes lanceolate; *upper lip of* the pale purple *corolla bearded.*—Waste places, around dwellings. (Naturalized from Europe.)

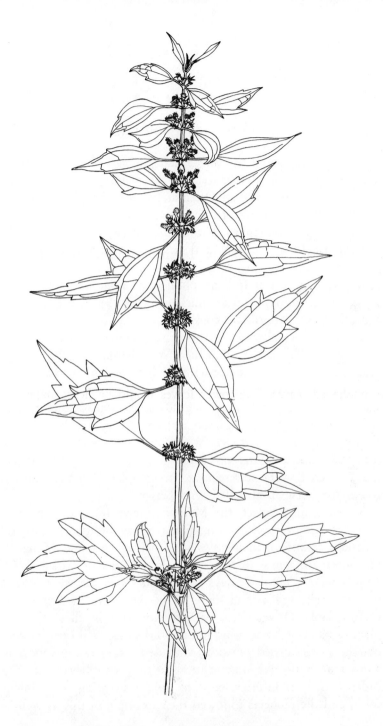

Fig. 70. The top of the Motherwort

HORSE MINT

[*Monarda fistulosa* L.]

HORSE MINT is one of the weeds that glories in hot weather and butterflies. Its spicy fragrance fills the country air of July and August as incense fills an Oriental palace. The plant does not lavish itself on the landscape as do so many other showy-flowered weeds, but it is sometimes seen in bunches large enough to make an armload, and it can be and is used in landscaping projects because of this growth habit. It is not so showy as its cousin the scarlet Oswego tea or Indian's plume (*Monarda didyma* L.) but it is just as fragrant and as much of a favorite of the bees and butterflies as is its brilliant cousin.

There are several weeds of this genus, *Monarda,* that are called horse mint. This is the one called Wild bergamont, Bee's balm and Oswego tea. It is the one with the lilac, or pink, or sometimes almost white flowers; the one seen along waysides, in fence rows and waste places, its flowers in heads at the top of square stems and at the ends of square branches; but its best mark of identification is its spicy fragrance. When one finds a mint with head-like flower clusters at the top of square stems, the long, two-lipped, lilac-colored flowers possessing a strong, spicy fragrance, he has found the horse mint called wild bergamont.

As one might expect, the Monardas have been and in some localities still are considered medicinal. They are used in teas and infusions because of their tastes and smells. The name, Monarda, was given the genus in honor of a French doctor, Nicolas Monardes, who in the early days of this country wrote several tracts on the merits and uses of some of the American medicinal and useful plants. Oswego tea was actually used in those days in place of tea, and horse mint, not the one called Wild bergamont, but the species called *punctata,* was used rather extensively only a few years ago—even now in rural districts—to relieve colic. The Indians used an infusion of it to induce sweating in incipient colds, and the pioneers followed their example in this as well as

Fɪɢ. 71. The Horse Mint or Wild Bergamot

in uses of several other plants. Monarda oil extracted from *fistulosa* as well as from other species of the genus was once used as a perfume for hair oil.

Fistulosa, the specific name, refers to the tube or pipelike corolla of the flower. Fistulosa is derived from the Latin word *fistulosa* meaning a pipe or tube.

TECHNICAL DESCRIPTION

Monarda L. Calyx 15-nerved, usually hairy in the throat. Corolla elongated, with a slightly expanded throat; lips linear or oblong, somewhat equal, the upper erect, entire or slightly notched, the lower spreading, 3-lobed at apex, its lateral lobes ovate and obtuse, the middle one narrower and slightly notched. Stamens elongated, ascending, inserted in the throat of the corolla.—Odorous erect herbs, with entire or toothed leaves, and large attractive flowers in a few verticels closely surrounded by bracts. (Dedicated to *Nicolas Monardes,* author of many tracts upon medicinal and other useful plants, especially those of the New World, in the latter half of the sixteenth century.)

Monarda fistulosa L. Branches more or less villous or hirsute, 0.5–1.5 m. high; leaves ovate-lanceolate, pubescent especially beneath, *the uppermost and outer bracts somewhat colored* (whitish or purplish); *calyx* slightly curved, *very hairy in the throat; corolla* 2.5–4 *cm. long, lilac or pink,* the upper lip very hairy.—Dry soil, New England to Colorado and Texas; often cultivated and mostly introduced northeastward.

BLACK NIGHTSHADE

[*Solanum nigrum* Tourn. L.]

THERE seems to be no real reason for this plant's having any of the names it has. Black nightshade, Deadly nightshade, Poison berry, and Garden berry are among the best known of its English names, and *Solanum nigrum* is its botanical name. The fruit of the plant is as nearly black as fruits get when they are ripe, and so it might be called the Black nightshade, but not the Deadly nightshade, or Poison berry, for it is neither deadly nor very poisonous, and Garden nightshade is not good either, for it is found more often in neglected places than in gardens.

Fig. 72. The Black Nightshade when it is
most difficult to identify

But why *nightshade?* What is there about the plant or the derivation of the word to suggest the name *nightshade?* Well, one of the family is Belladonna, the Deadly nightshade of Europe, and it was the potent juice of this plant that the witches brewed in the shades of night.

So nightshade is the name of the whole family, and a wonderful family it is. To it belong the potato, the tomato, tobacco, eggplant, ground cherries, and wonder berries as well as such weeds as the horse nettle, the buffalo bur, and the jimson weed. The family is known for its poisonous juices. Tobacco juice is the most potent of the lot. Pure nicotine extract is deadly. Atropine, a drug used in many ways but especially to dilate the pupils of eyes, can be extracted from the leaves of several of the nightshades, and the black nightshade yields this as well as solanidine, another useful drug.

The nightshade family, therefore, is a family of contradictions. It furnishes the best of food along with pleasure to millions of smokers, and yet the leaves of nearly every species in the family are poisonous to a greater or less degree.

The Black nightshade is not a bad weed, but it is one that should be known to all who work with plants. The drawing is of a flowering branch at the stage of growth when the plant is hardest to identify. At this stage the little white flowers and the deep-green ovate leaves are about the only identification marks. Later, with its small, loose bunches of small, green berries, and still later when these berries turn dark blue, there is no other weed like it. The ripe berries are said to be poisonous to children but non-poisonous to adults. Varieties of the Black nightshade have been selected for their large fruits and are grown in gardens as "Wonder berries," "Pie berries," "Garden huckleberries," and "Garden blueberries." Of course they are not the equal of the genuine blueberries as a pie fruit, but they make a good substitute for them.

TECHNICAL DESCRIPTION

Solanum (Tourn.) L. Calyx and wheel-shaped corolla 5–parted or 5–cleft (rarely 4–10–parted), the latter plaited in the bud, and valvate or induplicate. Stamens exserted; filaments very short; anthers converging around the style, opening at the tip by two pores or chinks. Berry usually 2-celled. Herbs, or shrubs in warm climates, the larger leaves often accompanied by a smaller lateral one; the peduncles also mostly lateral and extra-axillary.—A vast genus, chiefly in warmer regions. (Name of unknown derivation.)

Solanum nigrum L. Low, much-branched and often spreading, nearly glabrous; the stem rough on the angles; *leaves ovate, wavy-toothed; flowers* white, *in small umbel-like lateral clusters,* drooping; *calyx spreading;* filaments hairy; *berries* globular, *black.*—Shaded and rich open grounds; appearing as if introduced, but a cosmopolite. July–September.

HORSE NETTLE

[*Solanum carolinense* L.]

THE Horse nettle, Bull-nettle, Devil's tomato, Devil's potato, and Apple-of-Sodom are all names of this weed. The Devil's tomato is most fitting. It surely belongs to his satanic majesty and it is a little tomato.

The Horse nettle is the one weed that has thorns on it and bears a yellow tomatolike berry the size of a cherry. When in bloom it has a white or purplish white flower that reminds one of the tomato and potato flowers. It is a relative of these two plants, and for that reason, if for no other, it is an undesirable weed. Its leaves are eaten by the potato beetle, and thus this bad insect is kept alive and permitted to multiply whether potatoes are raised or not. Sometimes the farmer plants potatoes where he has never had them before, but when the shoots come through the ground he finds big, fat potato beetles have taken possession of them. He wonders why. The reason is that there were horse nettles in the field the year before. Some tomato and potato diseases are also most likely to be spread by this worthless weed.

The Horse nettle is found in gardens and cultivated fields more often than anywhere else, but it is able to grow in poor meadows, and in neglected fence rows. It is not a tall weed, and of course it cannot grow where other plants overshadow it. Moral

Fig. 73. The Horse Nettle in flower and fruit

—if you wish to rid your ground of Horse nettles grow good crops of anything that grows taller than the nettles do. Remember, too, that the weed is a perennial and that deep plowing will help in its eradication.

The meaning of the Latin name of the horse nettle, *Solanum carolinense,* is the Carolina nightshade.

TECHNICAL DESCRIPTION

See Black Nightshade for technical description of *Solanum* (Tourn.) L.

Solanum carolinense L. *Hirsute or roughish-pubescent with 4–8-rayed hairs; prickles stout, yellowish,* copious (rarely scanty); *leaves oblong* or ovate, obtusely sinuate-toothed or lobed or sinuate-pinnatifid; racemes simple, soon lateral; *calyx-lobes acuminate;* berry 1–1.5 cm. in diameter.—Prairies and plains, Missouri to Texas, and westward; adventive eastward.

JIMSON WEED

[*Datura stramonium* L.]

THE Jimson weed is at its best in hog lots. It delights in rich soil and in its own strong odor, which prevents the hogs from eating it. Under its wide-spreading branches the hogs make their trails, and above the trails the Jimson blooms and fruits in all its stinking glory. Of course it grows in other places than hog lots. It is almost as sure as the cocklebur to get into the cornfields of the lazy farmer, and it takes advantage of all poorly cultivated spots, if they are rich spots.

It is an easy weed to identify. Its odor alone will identify it. It takes but a single whiff of a crushed leaf to complete the education of the uninitiated. From that time on through life that sickening, never-to-be-forgotten odor will fill his nostrils every time he brushes against a Jimson weed.

Before this sad experience with the bruised leaf, the uninitiated may behold some beauty in the Jimson. Its broad green leaves and its white or purple trumpet-shaped flowers, which arise from the crotches of the forking branches, and the general shape of the plant itself, which is not unlike that of a spreading shade tree, are all beautiful enough until the odor of the leaves assails one's nostrils. Then he knows why the weed has been called Stinkweed, Stinkwort, Devil's trumpet, Devil's apples, Mad apple, and Thorn apple.

It is called Jimson weed as a corruption of Jamestown weed. It seems to have first appeared in America at Jamestown, Va. The Indians of that section of the country called it "Whiteman's weed," as it seemed to grow only near the white man's home.

Since it is an annual it is not a difficult weed to eradicate, even though a single plant will produce thousands of seeds. It grows so rank that only a few weeds can occupy any limited space. For this reason a lot may be filled with Jimson weeds and a few licks with a sharp hoe or scythe will leave it weedless. And if this is done before the seeds ripen it will clean up the place for some time to come.

The Jimson played a big part in the days of witches. There was a time in England when any one who permitted the weed to grow in his lots was in danger. The plant was the Devil's own in those days. As a weed eradicator the belief in witches was far more effective than modern legislation.

The Jimson is one of the best sources of atropine, a drug used for dilating the pupil of the eye. It is said that mountaineer girls who want to make their eyes shine squeeze a little of the juice of Jimson leaves into them. Even rubbing one's eyes with hands that have handled the weed will cause the pupils to dilate.

The botanical name, *Datura stramonium,* is meaningless so far as authorities on word derivations know. *Datura* seems to have come from the Hindu, perhaps the Sanskrit. Gray says "Altered from the Arabic name *Tatorah*." *Stramonium* is of unknown origin, but was at one time used as the specific name of another species of nightshade.

FIG. 74. The Jimson Weed, a relative of the tomato

TECHNICAL DESCRIPTION

Datura L. Calyx prismatic or cylindrical, 5-toothed, separating transversely above the base in fruit, the upper part falling away. Corolla funnel-form, with a large and spreading 5–10-toothed plaited border. Stigma 2-lipped. Capsule globular, prickly, 4-valved, 4-celled except near the 2-celled top. Seeds rather large, flat.—Rank weeds, narcotic-poisonous, with ovate leaves, and large showy flowers produced all summer and autumn on short peduncles in the forks of the branching stem. (Altered from the Arabic name, *Tatorah*.)

Datura stramonium L. Annual, *glabrous;* leaves ovate, sinuate-toothed or angled; *stem green; calyx prismatic; corolla white,* 7–9 cm. long, the border with 5 teeth; lower prickles of the capsule mostly shorter.—Waste grounds; a well-known ill-scented weed. (Naturalized from Asia?)

Another species, *Datura tatula,* is mostly taller than *stramonium,* has purple stems, and the corolla is pale violet purple.

MULLEIN

[*Verbascum thapsus* L.]

THE easiest weed in the world to identify is the mullein. It stands out among weeds like the crow among birds. It is the weed that any one would call "Jacob's Staff" or "Flannel-leaf." Those are two of its most descriptive names. Its stafflike stalk and the rosette of big, soft, flannel-like leaves at its base (there are smaller flannel-like leaves too, growing smaller all the way up the stalk) suggest to every one who sees the plant both "staff" and "flannel," so "Jacob's staff" and "Flannel-leaf" are names that are arrived at without any strain on the imagination of him who sees the plant for the first time.

There are several other common names that are more or less descriptive: Velvet leaf; Velvet dock; Torches; Hedge taper; Candlewicks (a name referring to the fact that the abundance of

FIG. 75. Common Mullein. No wonder its stalks
are called torches

hair on the leaves was scraped off and used by peasants to make their candlewicks); Big taper; Blanket leaf; Cow's or Clown's lungwort; Feltwort; Hare's beard; Shepherd's club; Peter's staff; Torchwort; Velvet plant; Adam's flannel; Old man's flannel; and Our lady's flannel. These are a few of them.

The name mullein is not descriptive in the least. It is a corruption of the Latin name Mulandrum, from which comes Melanders, meaning leprosy, and it refers to the fact that the mullein at one time was thought to be a remedy for leprosy. The weed is found all over the world, and from time immemorial has been among the medicinal plants. It produces a mild narcotic. Its tea has a sedative effect, and when asthma patients smoke the dried leaves they are said to find relief. But that is not why small country boys smoke mullein. The leaves look like tobacco leaves, and the smoke looks like tobacco smoke even if it does not smell like it, and so dried mullein leaves help the dreaming farm boy partially to attain, without maternal opposition, the stage he so longs to reach.

The mullein is not a bad weed. It is a biennial, and for this reason cannot establish itself in cultivated fields. It is found in neglected meadows and pasture lands and along fence rows that are not too much overgrown. The mullein must have room enough to spread its fine rosette of basal leaves if it is to erect its Jacob's staff. For him who can see it there is beauty in that stalk even in the winter time, but the most beautiful part of the mullein is its rosette of basal leaves. The rosette is made in the first year and is at its best in midwinter. It is then that a walk across bare fields is likely to bring one to a gentle, half-sodded slope on which are spread the soft green rosettes of the mullein. It is a thrilling experience that only he "who in the love of nature holds communion with her visible forms" can know.

The mullein is one of the easiest weeds to eradicate, for it cannot be crowded even by grass. If a good sod is kept in a pasture few if any mullein plants will appear, and if they do appear, a very few licks with a hoe any time between the formation of the rosette and the blooming stage will forever remove that generation of mullein plants from the field.

According to Gray's *Manual of Botany, Verbascum* is a modification of the Latin word *Barbascum,* which means mullein. *Thapsus* may be just a proper name, but probably comes from some resemblance to Thapsia, a genus of plants found in Europe.

TECHNICAL DESCRIPTION

Verbascum (Tourn.) L. Calyx 5-parted. Corolla 5-lobed, open or concave; the lobes broad and rounded, a little unequal. Style flattened at the apex. Capsule globular, many-seeded.—Tall and usually woolly biennial herbs; the leaves of the stem sessile, often decurrent. Flowers in large terminal spikes or racemes, ephemeral, in summer. (The ancient Latin name, altered from *Barbascum.*)

Verbascum thapsus L. *Densely woolly throughout; stem tall and stout, simple, winged by the decurrent bases of the* oblong acute *leaves; flowers yellow,* very rarely white, *in a prolonged and very dense cylindrical spike;* lower stamens usually beardless.—Fields, rocky or gravelly banks, etc., a common weed. (Naturalized from Europe.)

MOTH MULLEIN

[*Verbascum blattaria* L.]

To HIM who knows the common mullein, "Moth mullein" appears to be a misnomer. The common mullein is all stem, leaves, and fuzz; the Moth mullein is none of these, but is seen, if seen at all, because of its flowers. If more of its flowers were to bloom at once the moth mullein would be used as a decorative plant. As it is, with three or four blossoms out at any one time and with many globular seed boxes filled with seeds strung along the stem below the few flowers, the plant is conspicuously nothing but a weed. It feeds nothing except the worthless moths which visit its flowers in the evening and so give a reason for its name. How insignificant and worthless the plant is is attested by the fact it has only one common name. Yet it is the very kind of weed that should be treated in a weed book. It becomes conspicuous and almost beautiful when its first blossoms appear, and nearly every one who sees it then, for few know its name, asks, "What plant is that?" Its rosettes in early spring are beautiful, too; beautiful and worthless. The greens hunter is almost sure to find them and to think that here is something that ought to be good to eat. But they are not.

There is still another reason for knowing this worthless weed. It belongs to a wonderful family. The snapdragons are distant relatives of the moth mullein, as are also the gerardias, foxgloves, monkey flowers, and turtleheads. These all belong to the great Figwort family, the *Scrophulariaceæ*, the family with big lips, a family so named because in days of old it was thought that some of the group were effective remedies for scrofula. It contains a few beautiful plants, and a great number of just weeds; so the moth mullein is very much at home with most of its relatives.

The botanical name, *Verbascum blattaria*, is one such as Linnæus would construct. The generic part is a Latin modification of *Barbascum*, and the specific name is from the Latin *blatta*, which refers to things hidden or not easily seen. The *blattæ* are the night-flying moths; so even the botanical name of the plant means Moth mullein.

TECHNICAL DESCRIPTION

See Mullein for technical description of *Verbascum* (Tourn.) L.

Verbascum blattaria L. *Green and smoothish,* or somewhat glandular-pubescent above, *slender;* lower leaves petioled, oblong, doubly serrate, sometimes lyre-shaped, the upper partly clasping; *raceme loose, the pedicels longer than the fruit;* filaments all bearded with violet wool. —Roadsides and waste places, west Maine to Ontario, and southward, local. Corolla either yellow, or (in variety *albiflorum* Ktze.) white with a tinge of purple. (Naturalized from Europe.)

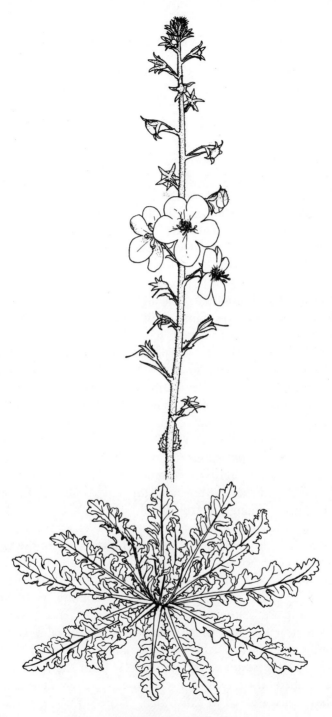

FIG. 76. The Moth Mullein and its
beautiful winter rosette

TRUMPET CREEPER

[*Tecoma radicans* Juss.]

THE Trumpet creeper is a very pretty flowering vine that can be used effectively in landscaping. Its large, orange-colored, trumpet-shaped flowers and its lacy leaves make it a most attractive plant when it is in bloom. When it fruits, its long, heavy, canoe-shaped pods are attractive also. These features of beauty, along with its ability to cling to walls, posts, trees, etc., cause it to fit nicely into some landscaping designs.

But it is a bad weed, a very bad weed. The numerous pods it bears are filled with hundreds of winged seeds, and one beautiful vine in some fence row or climbing up some old snag at the edge of a field will sow that field with seeds that seem never to fail to grow. The little plant roots deeply and when plowed out the part of the root left in the ground sends up a new shoot, while the part cut off finds itself buried in a moist place and begins its growth again. In a short time a field will be so overrun with these vines that nothing but cultivated crops can be grown in it. Cultivation will hold the vines in check, but only in check.

Because of this ugly feature the vine should never be used in landscaping near any farming community. There are other vines just as pretty and they do not have this ugly trait. No Trumpet creeper vine old enough to fruit should ever be allowed to do so if it is near a farm. The very first step in the eradication of this weed is to find the mother plants near the field where the work of eradication is needed. If the old plants are all cut off near the ground just after the pods have formed the chances are the roots will be killed and there will be no sprouting up of vines to take the place of the original vines. But the fence rows and wood lots near the field must be watched for several years, and the young vines in the fields must be plowed out, piled up and burned; and the sprouts that are sent up from the roots below must be pulled off or hoed off every time they show up, until the roots have been starved to exhaustion. If the sprouting vines are

Fig. 77. The Trumpet Creeper is
a beautiful bad weed

allowed to grow only a very few days, however, their leaves will feed the roots below and the fight will be greatly prolonged. So kill the old vines, plow out and burn up the young ones found in the field, and then follow up the good work by taking off the sprouts as fast as they come up until they cease to appear. You will find that a single year will practically rid the field of this pest.

The vine has several names: Trumpet creeper, Trumpet flower, Trumpet vine. Trumpet ash, Cross vine, and Cow-itch. In the South, Cow-itch seems to be the name most often used for the vine. Why the "itch" is attached to the name is not clear. Some say milk from cows feeding on the vine will cause an itching of the skin of those who drink the milk. Others say that when cows eat the vine an eruption occurs on the udders and those who milk the cows develop an itching skin. It is true that cows eat the weed. They are so fond of it that the Trumpet creeper vines in cow pastures are usually kept stripped of all tender shoots.

The weed is a member of the *Bignonia* family, and is a cousin of the catalpa tree. The generic name is said to be modern Latin derived from the Mexican name (originally Aztec) *tecomaxochitl,* which is the name of the tropical calabash tree. According to the *Oxford Dictionary,* Jussieu made a mistake when he attempted to place *Tecoma* in the family with the *tecomaxochitl.* It would seem that the only thing Antonine Laurent de Jussieu did exactly right when he named this genus was to use the pronounceable part of the unpronounceable word. He called his new genus *Tecoma,* pronounced tĕk'ō-ma by Gray's *Manual of Botany* and tĕ-kō'ma by Webster's *New International Dictionary.* The specific name, *radicans,* refers to the roots along the stem by which the vine clings to supports.

TECHNICAL DESCRIPTION

Tecoma Juss. Calyx bell-shaped, 5–toothed. Corolla funnel-form, 5–lobed, a little irregular. Stamens 4. Capsule 2–celled, with the partition at right angles to the convex valves. Seeds transversely winged.— Woody, with compound leaves, climbing by aerial rootlets. (Abridged from the Mexican name *tecomaxochitl.*)

Tecoma radicans (L.) Juss. Leaves pinnate; leaflets 9–11, ovate pointed, toothed; flowers corymbed; stamens not protruded beyond the tubular-funnel-form orange and scarlet corolla (6–8 cm. long); pod oblanceolate, 1–1.5 dm. long.—Moist soil, New Jersey to southeast Iowa, south to Florida and Texas; common in cultivation farther northward. August–September.

COMMON PLANTAIN

[*Plantago major* L.]

PLANTAIN, Common plantain, Greater plantain, Dooryard plantain, Birdseed plantain, Broadleaf plantain, and Roundleaf plantain are the names of a very common weed found in lawns and back yards throughout the United States. It is a perennial, and like the dandelion must be taken out by the roots if it once becomes established in a yard. Its leaves are so large and coarse that it cannot be disregarded as dandelions may be during the summer months. There is no beauty in its flowers either, and so it has not a single redeeming feature. But it does not root deeply, and is easily removed either by pulling when the ground is soft or by "spudding" at any time. The plant can stand a great deal of punishment and often grows in yards where the bluegrass has been injured either by tramping or by the close cropping of chickens and other domestic fowls. Goslings nearing the goose stage seem to delight in patches of plantain, especially when they are returning from their day's sports. They will gorge themselves on the tough leaves until their distended crops swing across their necks like goiters.

Common plantain is not likely to be found to any great extent in meadows or clover fields, and yet a very few plants in a clover field may supply enough seed to lessen the value of the clover-seed crop produced by that field. Since the plant is a perennial its flowering and fruiting do not occur before the second year, and for this reason it has never become a serious pest in cultivated fields or well-worked gardens.

If care is taken, plantain can be kept out of a lawn, for the seeds do not carry far, except on high winds or in whirlwinds. Any weed seed may be carried to considerable distance on high winds and especially in whirlwinds, and for this reason the appearance of a weed where it has not been seen before is not an unusual phenomenon. Of course a single plantain produces a great many seeds, and he who neglects to remove from his yard the first one he discovers there can expect to have more of these ugly plants in a short time, after his single plant has seeded. If the lawnmaker knows the plantain leaves when he sees them, and if he will take up the plants whenever he discovers them, it will not be very difficult for him to keep broadleaf plantain out of his yard. It will not be necessary for him to use weed-killing sprays on these plants. When there are enough of them to require weedkilling sprays the time has come to plow up the yard or meadow in preparation for reseeding.

TECHNICAL DESCRIPTION

Plantago (Tourn.) L. Calyx of 4 imbricated persistent sepals, mostly with dry membranaceous margins. Corolla salver-form or rotate, withering on the pod, the border 4-parted. Stamens 4, or rarely 2, in all or some flowers with long and weak exserted filaments, and fugacious 2-celled anthers. Ovary 2-celled, with 1-several ovules in each cell. Style and long hairy stigma single, filiform. Capsule 2-celled, 2-several-seeded, opening transversely, so that the top falls off like a lid and the loose partition (which bears the peltate seeds) falls away. Embryo straight, in fleshy albumen.—Leaves ribbed. Flowers whitish, small, in a bracted spike or head, raised on a naked scape. (The Latin name.)

Plantago major L. Smooth or rather hairy, sometimes roughish; *leaves thick and leathery*, 0.5–3 dm. long, *the blade from broad-elliptic to cordate-ovate*, undulate or more or less toothed, the broad petiole channeled; *scapes*, 1.5–9 dm. high, *commonly curved-ascending;* spike dense, obtuse, becoming 1–4 dm. long; sepals round-ovate or obovate; *capsule ovoid, circumscissile near the middle*, 8–18-seeded; *seeds angled, reticulated.*—Waysides and near dwellings, exceedingly common. —Sometimes with leafy-bracted scapes or with paniculate-branched inflorescences.

FIG. 78. The Common or Broad-leaved Plantain

BUCKHORN

[*Plantago lanceolata* L.]

BUCKHORN is the name most people who know this plant use when they speak of it, but it is also called Narrow-leafed plantain, Ripple-grass, Rib-grass, Ribwort, Buck plantain, and Buckhorn plantain. It is sometimes confused with Bracted plantain, an annual weed that resembles Buckhorn somewhat.

The Latin name of the species, *lanceolata,* is far more descriptive than any of the English names. *Plantago lanceolata* means the lance-leaved plantain, and that is what it is. Were its leaves made of steel they would make perfect pigmy lance heads.

Like all of the plantains, buckhorn has almost no redeeming feature. It is a perennial, and once it gets in it never goes out of its own accord. Its seeds are one of the worst contaminators of red cloverseed, and because of this the one redeeming feature the plant has may be pointed out. He who sows cloverseed containing buckhorn seeds is forced to turn under the crop he gets in order to rid himself of the weeds. This is the very best way to use a clover crop, and so the villain weed plays a hero's part.

Buckhorn is so villainous, however, that it deserves to be declared an outlaw by every State in the Union. If it is not found in every State that makes no difference. The weed has not been on this continent a great many years, and yet Muenscher can say of it, "Widespread throughout the United States and Canada."

The best way to rid a field or lawn of buckhorn is to plow the weeds under in May and keep the ground worked down and free from weeds until the last week of August or the first week in September (depending upon the latitude of the place), when the field or lawn is reseeded. A good sod should be obtained in this way, and if it is kept free of buckhorn for a year or two it is not likely that the outlaws will get in again.

TECHNICAL DESCRIPTION

The general description of the genus *Plantago* (Tourn.) L. is given in the sketch on Common plantain.

FIG. 79. Buckhorn, the clover
seed contaminator

Plantago lanceolata L. Mostly hairy; scape grooved-angled, at length much longer than the *lanceolate or lance-oblong leaves,* slender, 2–7 dm. high; spike dense, at first capitate, in age cylindrical; bracts and sepals scarious, brownish, *seeds 2, hollowed on the face.*—Very common in grassland. (Naturalized from Europe.)

BRACTED PLANTAIN

[*Plantago aristata* Michx.]

THE one plantain that has a semblance of virtue and just a hint of beauty is the smallest of the lot, the Bracted plantain. In different localities it is called Western buckhorn, Bristly buckhorn, Rat-tail plantain, and Western ripple-grass. The botanical name is *Plantago aristata,* which means the bristled plantain, but 'the English name, Bracted plantain, is quite as good as the scientific name if it would only stay put. That is the reason for scientific names. They are taken from the dead languages and so are the same in all languages and in all localities. English names are not the same as the German names, nor yet the Chinese names for the different plants and animals, but the scientific names are the same, and the botanists of every nation on the globe know what an American is talking about when he says *Plantago aristata.*

The Bracted plantain is one of the best plant indicators of sour soil. This is its special virtue. Where Bracted plantain grows to the exclusion of all other plants it is almost a hundred-to-one shot that the soil is sour. If the soil of such places contains the requisite constituents for plant growth the plantain may attain its normal height of from six to ten inches; but if the soil is poor as well as sour the plants may be so dwarfed that their massed appearance will resemble a short, thin grass. Then the stipes will be only three or four inches high, crowned with very short spikes of flowers and seeds. But in the nice eye of nature they are good weeds. They make seeds that will grow.

Good, sweet ground is the best eradicator of Bracted plantain. The plant is too small to compete with crop plants or with other weeds that are unhampered in their growth. Seed of the Bracted

FIG. 80. Bracted Plantain, the sour soil indicator

plantain contaminates cloverseed, but such cloverseed comes from fields that are not producing a perfect crop of clover. There are sour spots in the fields, or there are patches where the clover was frozen out or where the catch failed. It is in such patches that the Bracted plantain takes its stand, and from such patches its seed mingles with that of the clover at hulling time.

TECHNICAL DESCRIPTION

See Common Plantain for technical description of *Plantago* (Tourn.) L.

Plantago aristata Michx. Similar; *loosely hairy and green,* or becoming glabrous; the narrowly linear bracts 2–6 times as long as the flowers. —Dry plains and prairies, Illinois to Louisiana, and westward; naturalized in sterile soil eastward to the Atlantic.

BEDSTRAW

[*Galium aparine* L.]

WHEN one walks into some loose, trailing weeds and they all seem glad to be torn from their roots in order to ride away clinging to his trousers, he has made contact with one of the many Bedstraws. Gray's *Manual of Botany* lists twenty-five species of Galium (all bedstraws), but if the stems cling to the clothing the chances are that the find is one of two of them, *Galium aparine* or *Galium asprillum.* These are the two with the rough stems— the two that have this unique way of scattering their seeds. For of course those straws cling to the sheep's wool and the dog's hair just as they do to clothing, and sometimes they and their seeds are carried a long way by animals.

The species treated here, *aparine,* is one of the most widely spread and one of the easiest to identify. The six to eight leaves in a whorl around the stem, and the square or ribbed stems with their hooked spines, identify it. It is an annual and makes its appearance in fence rows and along roadsides. It often gets into mead-

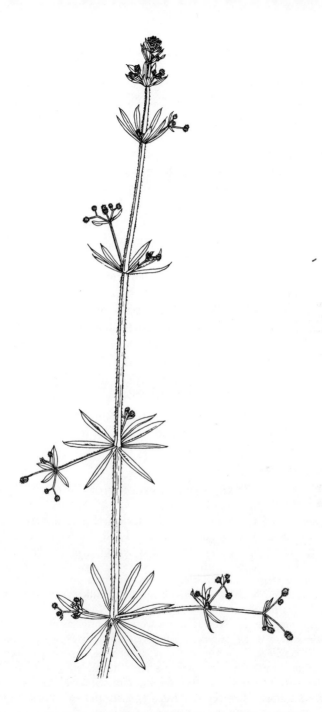

FIG. 81. Bedstraw with its
cleaving bristles

ows, too, but it never thrives in pastures, since nearly all grazing animals are fond of it. Even geese will keep it cleaned up where they are permitted to pasture on it. One of its common names is Goose-grass.

The fact that the weed is related to coffee may account for its being so relished as a forage plant. It is said that the seed of this particular bedstraw is one of the very best substitutes for coffee.

This species of bedstraw grows in England, and from there come several of its common names. It is called Cleavers, Catchweed, Hay ruff, Mutton-chops, Scratch-grass, Grip-grass, Goosegrass, and Robin-run-in-the-hedge. The botanical name comes from the Greek, and the generic part, *Galium,* means milk, since some of the species of bedstraw have been used, and probably are now used, to curdle milk for cheese making. This species, *aparine,* because the straws will cling together, has been used to make coarse sieves for straining milk. The specific name, *aparine,* also from the Greek, means to seize. So a liberal translation of *Galium aparine* would be something like "the milkweed that seizes."

The roots of this weed make a red dye, and a wash made by boiling the stems is said to be good for sunburn and freckles. It is not a bad weed but an interesting one, and for this reason every one should make its acquaintance.

TECHNICAL DESCRIPTION

Galium L. Calyx-teeth obsolete. Corolla wheel-shaped, valvate in the bud. Stamens 4, rarely 3, short. Styles 2. Fruit dry or fleshy, globular, twin, separating when ripe into the 2 seedlike indehiscent 1–seeded carpels.—Slender herbs, with small cymose flowers (produced in summer), square stems, and whorled leaves, the roots often containing red coloring matter. (Name from Greek meaning milk, which some species are used to curdle.)

Galium aparine L. Annuals. Flowers in long ascending axillary peduncles. *Stem weak and reclining, bristle-prickly backward,* hairy at the joints; *leaves about 8 in a whorl,* lanceolate, tapering to the base, short-pointed, rough on the margins and the midrib, 2.5–7 cm. long; *peduncles 1–3–flowered;* flowers white; fruit bristly, 3–4 mm. in diameter.—Seashores, Quebec, Florida, and in rich or shaded ground inland; perhaps sometimes introduced. (Eurasia.)

POOR JOE

[Diodia teres Walt.]

POOR JOE is a little, sprangly madder that grows in the poorest of cultivated ground, from New England to the Rocky Mountains and south: especially south. Poor Joe is supposed to have originated in Mexico along with some other weedy characters, and along with them to have established himself in the United States wherever the more aristocratic emigrants and natives refused to go.

The weed is insignificant and scarce in rich soil areas, but it seems to rejoice and flourish in difficult situations. A hot wheat or oat stubble field in the South, where either the wheat or the oats were so thin and short that the ground had to be shaved to recover the grain, will become as green with the wiry stems and harsh little leaves of Poor Joe as will the rich stubble fields of the North with ragweeds. Like the ragweed, it should be used as a soil builder. It should be plowed into the soil, and such a baptism ought to give to it another name: "Helpful Joe" or "Restoration Joseph."

Poor Joe has three other common names: Button weed, Poverty weed, and Poor-land weed. The name "Button weed" refers to the shape of its fruits, which are hard and somewhat like very thick little buttons. Button weed is a very good name. Its other names refer to the weed's favorite habitat: poor soil.

Poor Joe can and does grow in good soil when given a chance. The only reason it is seldom seen in such places is because it is smothered out there by the taller and better-leaved weeds. Where the strong-growing weeds are kept down, however, Poor Joe gets in and becomes rather robust, again taking advantage of the situation.

The generic name, *Diodia,* is derived from a Greek word meaning thoroughfare. Some diodias were seen by the wayside. Poor Joe may be seen there if the ground of the wayside is poor enough to keep out the usual wayside weeds, but Poor Joe was not respon-

FIG. 82. Poor Joe or Button Weed

sible for the name, Diodia. A European plant, a cousin of Joe's, was probably in the mind of Linnæus when he named the genus. Thomas Walter had in mind the bristle-like stipules of its leaves when he gave the species the name *teres*. The word carries the meaning of round and thin or slender, and the stipule hairs near the top of the plant are just that: round and slender.

TECHNICAL DESCRIPTION

Diodia (Gronov.) L. Calyx-teeth 2–5, often unequal. Fruit 2 (rarely 3)–celled, the crustaceous carpels into which it splits all closed and indehiscent. Flowers 1–3 in each axil.—Resembling *Spermacoce*. Flowering all summer. (Name from the Greek word meaning *a thoroughfare;* the species often growing by the wayside.)

Diodia teres Walt. Hairy or minutely pubescent annual; stem spreading, 1–8 dm. long, nearly terete; leaves linear-lanceolate, closely sessile, rigid; *corolla funnel-form, 4–6* mm. long, whitish, with short lobes, not exceeding the long bristles of the stipules; *style undivided;* fruit obovoid-turbinate, *not furrowed,* crowned with 4 short calyx-teeth.—Sandy shores and barrens, Connecticut to Florida; and from Ohio to Kansas, and southward. (Mexico, West Indies.)

BUCK BRUSH

[*Symphoricarpos orbiculatus* Moench.]

Where it grows, and it grows in far too many places, there is no worse weed in pasture lands than the little bush honeysuckle known as Buck brush, Coral-berry, or Indian currant. It is so beautiful because of its habit of growth and its profusion of red berries that it is often used as a decorative plant, but it will take over pasture fields, and the farm boy who has had to spend hot days grubbing out the pest is not likely to become very enthusiastic when its beauties are extolled.

The beautiful berry clusters which give the weed its place

among decorative plants hold only one of its means of propagation. The berries furnish a winter diet for birds which sow the seeds in that beautiful, mutually helpful way that nature so often uses. You scratch my back and I'll scratch yours. Or, you sow my seeds and I will grow you more fruit to eat, and the cows that won't sow my seeds can look elsewhere for their grass. That is nature in the raw.

So the seeds are sown, but if the birds go on a strike, or if for some reason the seeds are not scattered, Buck brush has another string to its bow, and it is actually a string. The viny nature of some of the weed's relatives shows up in this plant, and out from its base crawls a long vinelike shoot that takes root at several places along its creeping stem. Wherever this stem roots a new plant will spring up, and by this means every clump of Buck brush increases in size with every growing season.

As a decorative plant in lawns where the lawn mower is used Buck brush can be kept under control and made to serve as a low hedge or screen quite as well as barberry does. Selected varieties of the wild plant are sold under the names *Symphoricarpos rubra* or *S. orbiculatus,* but they are just Buck brush with redder or larger berry clusters than the ones usually seen in the wild. Where the plant grows in abundance in pastures any one who cares to do so can make just such selections as these. He has only to look until he finds a clump that produces fine berry clusters and has a pleasing habit of growth. Such plants will show up even better when they are transplanted to a landscaped yard.

Buck brush is a name applied to any low brush that sheep or deer will feed upon, but most of the country folk of the Mississippi Valley know this particular plant as Buck brush. The other two common names, Indian currant and Coral-berry, are so descriptive that one who sees the plant in winter time knows it by either of these names. *Symphoricarpos,* the generic name, is made up from two Greek words, the first meaning to bear together, and the second meaning fruit, that is, fruit borne in bunches. *Orbiculatus* refers to the round leaves. The round-leafed plant that bears its fruit in clusters is a free translation of the scientific name.

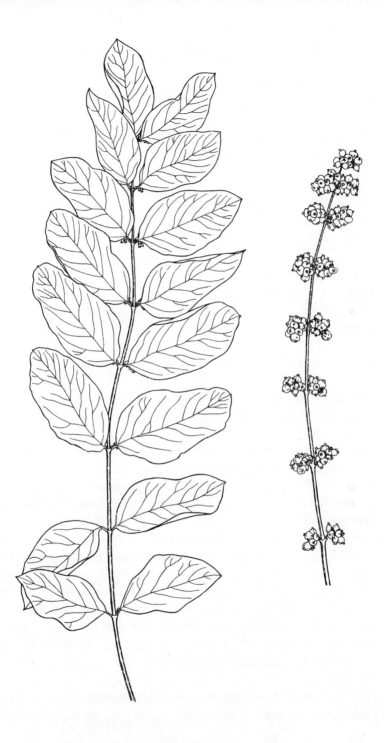

FIG. 83. The Buck brush in flower and fruit

TECHNICAL DESCRIPTION

Symphoricarpos (Dill.) Ludwig. Snowberry. Calyx-teeth short, persistent. Corolla bell-shaped, regularly 4–5-lobed, with as many short stamens inserted into the throat. Berry 4-celled, 2-seeded. Seeds bony.— Low and branching upright shrubs, with oval short-petioled leaves, which are usually downy underneath and entire, or wavy-toothed or lobed on the young shoots. Flowers white, tinged with rose-color, in close short spikes or clusters. (Name composed of the Greek word meaning *to bear together* and another Greek word meaning *fruit;* from the clustered berries.)

Symphoricarpos orbiculatus Moench. Flowers in the axils of nearly all the leaves; corolla sparingly bearded; berries small. (*S. vulgaris* Michx.; *S. Symphoricarpos* Mac M.)—Rocky banks, New York to Dakota, south to Georgia and Texas; escaping from cultivation eastward. July.

IRONWEED

[*Vernonia altissima* Nutt.]

WHEN Ironweed is mentioned to a farmer he thinks of one of three different weeds. It may be this one (or one of its near relatives), its straight stem topped with purple flowers; or it may be one or more of the vervains; or he may think of the late-flowering thoroughwort, the dirty-white eupatorium that often grows among the purple-topped vernonias. Whatever weed it may be that dominates the farmer's mind when ironweed is mentioned it deserves to be called "ironweed," for the hard, woody stems of all of these plants are ironlike and persist throughout the winter.

But this particular weed, *Vernonia altissima,* seems to have no other name, unless it be the Hoosier corruption, Urnweed, and perhaps the greater part of those who say "Ironweed," have in mind some one of the eight species of vernonias we have in America.

Ironweeds are very conspicuous in late summer and in winter landscapes. This species, *altissima,* is so tall and its purple-flowered

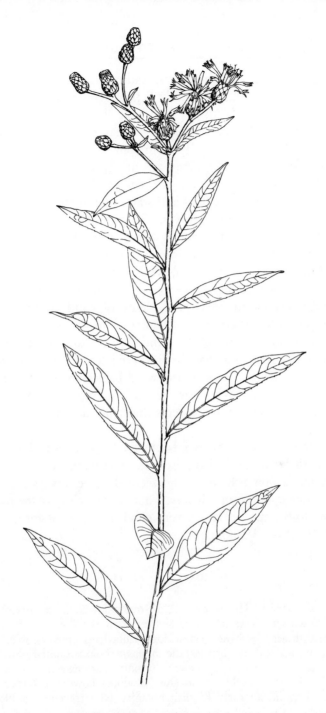

FIG. 84. A very small specimen
of the Ironweed

heads in a "cyme, large, widely spreading, rather loose," are so attractive in late August and early September that any one travelling through the Mississippi Valley at that time of the year is almost sure to see and admire it. It may become as much as seven feet tall, but it is usually from four to five feet in height, and its straight but leafy stems are nearly as attractive as the purple flower clusters at the top of them.

The ironweed is a perennial and starts its growth as most perennials do, with a rosette of leaves. The stem that grows up from the rosette develops slowly, but by the first of August the flower heads are all set and by the middle of the month the purple flowers will be showing. After the flowers come the fuzzy seeds that have bristly wings enough to carry them, but are so tightly fixed on their receptacles as to require plenty of wind-tossing of the resilient stems to loosen and pitch them into the air. The wing surface is not downy enough for gentle breezes to carry the seeds, so Nature has seen to it that the ironweed seeds start their journeys only on rough days. The ironweed babies, then, ride off on iron-testing gales, and wherever they fall, on hillside or in slough, they produce strong, straight stems that truly deserve their common name.

The genus to which this weed belongs was named in honor of the English botanist, William Vernon. The name was given by Johann D. C. von Schreber, who seemed to be always looking for some one to honor in this way. Thomas Nuttall gave the name *altissima* to the species. It is a very good name, for it means the tall species of the Vernonias.

TECHNICAL DESCRIPTION

Vernonia Schreb. Heads discoid, 15–many-flowered, in corymbose cymes; flowers perfect; involucre shorter than the flowers, of much imbricated bracts. Achenes cylindrical, ribbed; pappus double, the outer of minute scalelike bristles, the inner of copious capillary bristles. —Perennial herbs, with leafy stems, alternate acuminate or very acute serrate leaves and mostly purple (rarely white) flowers. (Named for *William Vernon,* an early English botanist, who travelled in North America.)

Vernonia altissima Nutt. Usually tall (1–2 or more m. high); *leaves lance-oblong,* acuminate, *spreading, smooth or merely puberulent beneath; cyme large, widely spreading, rather loose;* heads about 25-flowered; involucral bracts closely appressed, ovate, acute, obtuse, or cuspidate, mostly purple-tinged; flowers red-purple.—Rich soil of prairies, etc., New York to Michigan, Missouri, and southward; also sporadic northeastward.

LATE-FLOWERING THOROUGHWORT

[*Eupatorium serotinum* Michx.]

HERE is a weed that is found in overgrazed pastures and in the waste places of most of the States east of the Rocky Mountains, and yet few who know it have a name for it. It is called ironweed—if called anything—by the farmers, and it deserves the name quite as much as the Vernonias, those purple-topped weeds that often share space with this near relative of the white snakeroot. And that is one of the reasons for including the Late-flowering thoroughwort among the weeds of this book: it is sometimes mistaken for the White snakeroot. There is some resemblance between the two weeds, but the dirty-white flowers of the Thoroughwort and its heavier and narrower leaves are easily distinguished from the snow-white flowers and the rather broad, thin leaves of the Snakeroot, especially when the two plants are brought together for comparison.

Because of its late-blooming and its dirty-white flowers the weed escapes the notice of the casual observer, even though it is a tall weed, growing from three to six feet in height. It is a perennial, and although it blooms late its worthless shoots are absorbing sun and space from early spring on through the growing season, and its woody stems sway in the winds of the winter to scatter its bristle-winged seeds. It is a real weed, good for nothing, but filled with vigor and persistence. It is a near relative of the Joe Pye weed, whose crown of pink blossoms when seen in a mass pleases the beholder. It is just as closely related to Boneset, a weed famous for its place in Indian and home remedies; and the White snakeroot, the villain of the Eupatoriums, is another near relative. But it has neither the virtues nor the vices of its relatives. It is just a dull, worthless weed, or it would be if it were not for the fact that its flowers are fragrant and so attract nectar-hunting insects.

When a field becomes infested with the Late-flowering thoroughwort, and with such other weeds as the yarrow, the vervains, and the ironweeds, it should be plowed soon enough in the season to put all of their succulent leaves and stems down among the soil-inhabiting bacteria. That is the only way to use worthless weeds. They become valuable soil builders when so used.

The Late-flowering thoroughwort is a composite with very small flowers in very small heads, just as are all of the Eupatoriums and goldenrods. Daisies and sunflowers are composites, too, but their small flowers are collected in big heads. Which is the better type for seed production and dissemination may be a mooted question, but Nature does not seem to be concerned about it.

The name of the genus was given in honor of the Greek physician, *Eupator Mithridates,* whom Linnæus thought used one of the species of the genus in medicine. François André Michaux named the species *serotinum,* which means late-blooming. The word is derived from the Latin word *serotinus,* meaning to develop late or slowly.

TECHNICAL DESCRIPTION

Eupatorium (Tourn.) L. Heads discoid, 3–many-flowered; flowers perfect. Involucre cylindrical or bell-shaped, of more than 4 bracts. Receptacle flat or conical, naked. Corolla 5-toothed. Achenes 5-angled; pappus a single row of slender capillary barely roughish bristles.— Erect perennial herbs, often sprinkled with bitter resinous dots, with generally corymbose heads of white, bluish, or purple blossoms, appearing near the close of summer. (Dedicated to *Eupator Mithridates,* who is said to have used a species of the genus in medicine.)

Eupatorium serotinum Michx. Stem pulverulent-pubescent, bushy-branched, 1–2 m. high; leaves ovate-lanceolate, tapering to a point, triple-nerved and veiny, coarsely serrate, 0.5–1.5 dm. long; involucre very pubescent.—Alluvial ground, Maryland to Minnesota, eastern Kansas, and southward.

FIG. 85. The Late-flowering Thoroughwort

WHITE SNAKEROOT

[*Eupatorium urticæfolium* Reichard]

ONE OF the worst of weeds is the White snakeroot. It is the weed that causes milk sickness, and in dry seasons it so often becomes a menace that pictures of it are posted by Federal and State governments throughout its range; and that range is from the entire eastern coast of North America almost to the edge of the Rocky Mountains.

With all of the publicity it gets it would seem scarcely necessary to include White snakeroot in this book of weeds, but in spite of the many pictures and printed warnings people who have cows in pastures seldom know the weed when they see it. The fact that cattle do not like the taste of it and never eat it except when there is little else left to eat saves the lives of many who drink the milk from pasturing cows.

White snakeroot can be found in nearly every old pasture within its range. It is especially fond of shaded grasslands and of open wood lots. If a cow feeding in such places takes the "trembles" she has been eating White snakeroot, and if a child or any one else who has drunk that cow's milk becomes restless, weak, and languid, a doctor should be called, for milk sickness is not gentle with its victims. Later vomiting will set in, the patient's breathing will become labored, and there will be a characteristic fetid odor of the breath. The progress of the disease is said to be very much like that of typhoid fever. In severe cases the patient soon becomes delirious, goes into a coma and dies.

The white flower heads (the weed is a composite, and the flowers are made up of heads of many small flowers) and the long-stemmed (petioled), opposite, thin leaves ought to be enough to identify it when it is in bloom. And everybody who drinks milk ought to know the plant, even though he may never own or operate a dairy. He ought to look at pictures of White snakeroot and search for it in likely habitats of the late summer until he is sure he has found it and knows it whenever he sees it.

Fig. 86. The White Snakeroot in all
of its treacherous beauty

Then he can point it out to his dairyman and lay down this ultimatum: "You either grub out every one of those weeds or I and all my relations will cease buying milk from you." There is no reason why this menace to health cannot be exterminated. Native though it is, and possessing some medicinal properties though it does, it belongs to the category of malaria-carrying mosquitoes and copperhead snakes, and it is everybody's business to help in the extermination of such things.

There is, perhaps, only one other weed with which White snakeroot might be confused, and it is another Eupatorium known as the Late-flowering thoroughwort. It is a much taller weed than the snakeroot and its flowers are dull white instead of snow-white as are those of the White snakeroot.

This weed, White snakeroot, is also called White sanicle, Indian sanicle, Deerwort, Richweed, and Squaw-weed. It is a perennial and so has to be taken out roots and all if it is to be destroyed. It propagates by seeds only, but preventing its seeding once does not prevent its seeding again the next year if the roots remain in the ground, nor does it prevent the old plant from poisoning the old cow.

Much that has been said thus far has been said for the purpose of stirring the reader to action. The truth is that the well-fed cows of any dairyman who is permitted to sell milk will not eat White snakeroot. But other stock—horses, sheep, and hogs—will eat it, and it affects those animals in the same way that it does cattle. It is to be remembered, too, that all of the bovines that pasture are not milch cows. Calves, steers, bulls, and dry cows are often found in poor pastures where White snakeroot is abundant, and sometimes they eat the hateful weed and die or become worthless forever after. So there are still reasons for the eradication of the White snakeroot.

TECHNICAL DESCRIPTION

See Late-Flowering Thoroughwort for technical description of *Eupatorium* (Tourn.) L.

Eupatorium urticæfolium Reichard. Smooth, branching, 0.5–1 m.

high; *leaves* broadly ovate, *pointed,* coarsely and *sharply toothed, long-petioled, thin,* 7–12 cm. long; corymbs compound. (*E. ageratoides* L.) Rich woods, not rare. Var. *villicaule* Fernald. Stems and petioles viscid-villous.—Pennsylvania (Heller) to Virginia (Curtiss).

GOLDENROD

[*Solidago canadensis* L.]

HE IS blind indeed who does not know goldenrod, but he is a taxonomist if he knows all the goldenrods. *Gray's Manual* lists fifty-six species and varieties. Even the casual observer will notice that there are some that bloom early and some that bloom late, some that are short and some that are tall, some that are beautiful and some that are rather coarse and ugly. All of these differences are used in naming the different species, and all of the different species are just weeds.

The species treated here is one that is common from Kentucky northward, and it is easily identified. It is the straight, almost smooth plant with leaves that are lance-shaped and bear three principal veins. The flowers look very much like those of the other species, except, as the drawing shows, the cluster is more graceful than some of the others are. The flowers are really made up of numerous flower heads, with many little flowers in each head. The goldenrods are composites. They are all perennials, which accounts for their being seen almost entirely in waste places and along the waysides. They are scarcely ever seen in cultivated fields except when the fields have been allowed to lie fallow for a year, but they do get into meadows and hayfields, especially when such fields lie undisturbed for a season.

The goldenrods are truly weeds of the wayside, with emphasis on the "weeds." Aside from the beauty of some of the species, which has caused them to be adopted as State flowers in several States, the goldenrods have not a single commendable character, and they do have at least one very undesirable weedy trait. They are among the generators of hay fever. The "wondrous days of green and gold" become horrible days for some people when the goldenrods come on the scene.

Fig. 87. One of the many Goldenrods

The name *Solidago* comes from the Latin *solidare,* meaning to join or make whole. Some one or more of the species had the reputation, once upon a time, of possessing healing qualities, and the name refers to the goldenrod's ability to heal wounds. *Canadensis,* of course, refers to Canada as its favorite habitat. It is the Canadian goldenrod. It is also called the Rock goldenrod.

TECHNICAL DESCRIPTION

Solidago L. Heads few-many-flowered, radiate; the rays 1-16, pistillate. Bracts of the involucre appressed, destitute of herbaceous tips. Receptacle small, not chaffy. Achenes many-ribbed, nearly terete; pappus simple, of equal capillary bristles.—Perennial herbs, with mostly wandlike stems and sessile or nearly sessile, never heart-shaped stem-leaves. Heads small, racemed or clustered; flowers both of the disk and ray yellow. Closely related species tending to hybridize freely. (Name from *solidare,* to join, or make whole, in allusion to reputed vulnerary qualities.)

Solidago canadensis L. Stem rather slender, 0.3-1.5 m. high, *glabrous at least below,* often minutely pubescent above; *leaves narrowly lanceolate, thin, glabrous above, minutely pubescent on the nerves beneath, mostly sharp-serrate,* the middle ones 6-13 cm. long, 5-18 mm. wide; heads tiny, crowded in recurved racemes and forming *dense, broadly pyramidal panicles; pedicels strongly pilose;* involucral bracts linear, mostly attenuate, greenish-straw-color.—Thickets and rich open soil, Newfoundland to North Dakota, south to West Virginia and Kentucky. July–September.

DAISY FLEABANE

[*Erigeron annuus* (L.) Pers.]

THERE are two daisy fleabanes that look so much alike and behave so similarly that any one but a botanist may be forgiven if he calls *Annuus, Ramosus* or *Ramosus, Annuus.* They are both called White-top and White weed and both deserve both of these names. *Annuus* is sometimes called Sweet scabious, and *Ramosus* is the "rough daisy fleabane" in some localities.

The weeds are twinlike in many respects. They are both annuals or winter biennials, they both grow to about the same height, their flowers (flower heads) are about the same size, their leaves are much the same shape and size; but there are a few differences. There are more branches on *Ramosus* than on *Annuus.* That is what *Ramosus* means: branched or twigged. And *Ramosus* has a thinner stalk, and flowers that are a trifle smaller than those of *Annuus.*

Both of the weeds are lovers of hayfields and many a crop of hay is spoiled by them. Their dried leaves, which resolve themselves into a dusty powder on the least provocation, are said to chase fleas, but there are no fleas in hay, and fleabane leaf dust is not exactly what the stock feeder wants to feed his stock.

The best way to fight these weeds is to crowd them out with good vigorous sods. If the white-tops become more in evidence than the timothy or clover in the field—and it does not take a great many fleabane plants to make such a showing—it is time to plow under the mixture of hay and weeds. The soil will be enriched thereby, and after a cultivated crop or two the field can be expected to produce a sod that will keep out the fleabanes.

Erigeron (pronounced ė-rĭj′ ēr-ŏn), the generic name, is from two Greek words, *eri* meaning early, and *geron* meaning old man. The name refers to the gray appearance of some of the species. Freely translated the word means "soon becoming old," and such a translation is very descriptive of most of the species of the genus and especially of these two, *Annuus* and *Ramosus,* for they are experts in ripening their seeds in the shortest possible time after their flowers open.

Erigeron annuus, the annual erigeron, may be seen in almost any hayfield and along the roadsides in all parts of the United States and Canada wherever hay is grown. It is in full bloom in most places by the first of June, but blooming specimens can be found throughout the summer. The white or pink fringe made by the ray flowers around the yellow disk flowers that fill the center of each flower head make very pretty daisies. The flowers are small in comparison with those of the ox-eye daisy, but they are certainly daisylike, and "daisy fleabane" is a very descriptive name for the weed.

FIG. 88. The Daisy Fleabane,
a meadow weed

TECHNICAL DESCRIPTION

For the technical description of the genus *Erigeron* L. see technical description of the Horsetail fleabane.

Erigeron annuus (L.) Pers. Stem stout, 2–15 dm. high, branched, *beset with spreading hairs; leaves coarsely and sharply toothed; the lowest ovate,* tapering into a margined petiole; the upper ovate-lance-olate, acute and entire at both ends; heads corymbed; rays white, tinged with purple, not twice the length of the bristly involucre.—Fields and waste places; a very common weed. June–October. (Naturalized from Europe.)

HORSETAIL FLEABANE

[*Erigeron canadensis* L.]

THERE are few weeds more universally common than the Horse-tail fleabane, and yet even the farmers who see it on every hand seldom have a name for it. It is a weed with a lackluster personality. The flowers (actually flower-heads; the plant is a composite) and the seed-filled heads that follow the flowers are so little and inconspicuous that even though the plant may be six feet tall there is nothing about it to attract attention. True, it has a straight-as-an-arrow stem with simple, lance-shaped or paddle-shaped leaves scattered along it from the ground up to its muletail brush of flower and seed branches at the top. Such a top is called a panicle, but this is an unusual panicle. It is the one touch of individuality in the entire make-up of the weed. This panicle has caused those who have noticed it to call the weed Horsetail, Muletail, and Colt's-tail. For other reasons the weed has been called Fleabane, Fleawort, Bitterweed (it is bitter), Hogweed, Prideweed, and Bloodstanch.

The name, Fleabane, is no more applicable to this plant than it is to several other species of weeds. The scent of the fleabanes is supposed to be obnoxious to the flea's sense of smell. This particular plant does produce an oil which reacts very much like the oil of turpentine; so it may be that the fleas absent themselves

FIG. 89. A short-topped specimen of
Horsetail Fleabane

from the oil-bearing leaves of the Horsetail fleabane by leaps and bounds. The oil, along with some other medicinal properties that are found in the leaves of the weed, have placed it in Pharmacopœia. The name, Bloodstanch, attests the fact that extracts from the leaves and flowers arrest hemorrhages from the lungs and alimentary tract (according to herbals). At any rate the oil of fleabane is sold on the market and the leaves act as an astringent.

The weed is an all-American weed, but it is said to be found now in all parts of the world. It is an annual, and it springs up in great masses in wheat fields and other grain fields after the harvest, just as the ragweed does. And it can be used just as the ragweed can for green manure. Plowed under before it blooms it makes as good fertilizer as any of the weeds, and this is the only sensible way to keep it under control. In truth the weed does not need much controlling. There is nothing very obnoxious about it except that it will take places left open to it. Few hay-fever patients are bothered by this fleabane, and it is not aggressive enough to crowd out or shade out any other well-established plant. It is just a common, easy-going weed with a bitter juice and with oil enough in its make-up to give it a place in Pharmacopœia.

TECHNICAL DESCRIPTION

Erigeron L. Heads many-flowered, radiate, mostly flat or hemispherical; the narrow rays very numerous, pistillate. Involucral bracts narrow, equal, and little imbricated, never coriaceous, neither foliaceous nor green-tipped. Receptacle flat or convex, naked. Achenes flattened, usually pubescent and 2-nerved; pappus a single row of capillary bristles, with minuter ones intermixed, or with a distinct short outer pappus of little bristles or chaffy scales.—Herbs with entire or toothed and generally sessile leaves, and solitary or corymbed naked-pedunculate heads. Disk yellow; rays white, pink or purple. (The ancient name presumably of a *Senecio,* from two Greek words *spring* and *an old man,* suggested by the hoariness of some of the vernal species.)

Erigeron canadensis L. Bristly-hairy; *stem erect, wandlike,* 0.1–3 m. high; leaves linear, mostly entire, the radical cut-lobed; heads very numerous and small, cylindrical, *panicled.* (*Leptilon* Britton.) Waste places, etc., a common weed, now widely diffused over the world. July–October.—Ligule of ray-flowers much shorter than the tube, white.

HORSE WEED

[*Ambrosia trifida* L.]

THE Horse weed, or Giant ragweed, is another Ambrosia; *Ambrosia trifida,* the botanist calls it. Like the common Ragweed it does not deserve the name Ambrosia, but it does deserve the name Horse weed. Horses like it. The name Horse weed, however, probably refers to its size rather than to its being eaten by the horse. It is truly the Giant ragweed. In rich bottom-lands it will attain heights of from ten to fifteen feet with stems as hard as wood and as much as two inches thick at the base. The long, straight, dried stems are often used as javelins by country boys in their play. There is only one other weed that is likely to have a larger and harder stem than the Horse weed, and that is the Sunflower.

The Horse weed, like the Cocklebur, is a good indicator of soils. It grows very rank in good soils and is seldom noticed, if it grows at all, in poor soils. In poor soils it is crowded out by weeds better adapted to such places. It requires plenty of moisture and nitrogen to make its growth, but when it has these requirements few weeds can compete with it. A Horse-weed patch usually contains nothing but Horse weeds.

Of course its pollen is a menace to hay-fever subjects. A farmer friend of the author declares he did not know what hay fever was until he, as a young man, plowed under a big patch of Horse weeds that were in full bloom. That was a long time ago, but every year since then when the Horse weeds bloom he remembers the curse those stately weeds silently placed upon him as their pollen-loaded tops disappeared beneath the good rich soil his sulky plow was turning.

Like all rank weeds, Horse weeds make good fertilizer when plowed under soon enough. They should never be permitted to bloom for two reasons: first, they become too woody by that time, and second, their pollen is a menace to him who has never known hay fever.

Fig. 90. The Horse Weed

The weed is not hard to identify after it has become fully grown, and by the botanist it is easily identified at any time. The name *Ambrosia trifida* refers to its leaf: the three-parted-leafed Ambrosia. The *artemisiifolia* of the true ragweed refers to the leaf also. It is the ragweed with the leaf like that of wormwood (*Artemisia*). There is no other weed that has a leaf like that of the Horse weed. It has three large lobes, and it is rough.

Look for this weed in rich ground, in the garden, in lots, and especially in bottom fields where the plow failed to do its duty. It is a noble weed, if there is such a thing, and every one who works with soil should make its acquaintance.

TECHNICAL DESCRIPTION

The technical description of the genus *Ambrosia* (Tourn.) L. is given in the description of Ragweed.

Ambrosia trifida L. Stem stout, 1–6 m. high, rough-hairy, as are the large deeply 3-lobed leaves, the lobes oval-lanceolate and serrate; petioles margined; fruit obovoid, 5–6-ribbed and tubercled.—Rich soil, common westward and southward, much less so northeastward.

RAGWEED

[*Ambrosia artemisiifolia* L.]

THERE are at least three species of Ragweeds but the raggedest of the three, *Ambrosia artemisiifolia,* is the one that most deserves the name and is the one usually in mind when Ragweeds are mentioned.

That a great botanist should jest and name a genus of such plants as this *Ambrosia* (food of the gods), may show a sense of humor, but it certainly does not show good taste. This weed is too bitter to be eaten by anything but a cow, and she resorts to it only after the grasses are gone. Of course it spoils her milk. It is not quite as bad as wild garlic as a milk flavor, but it is a close second to the sneeze weed, a weed that makes milk taste like quinine.

This weed can readily be identified by the picture. Its lacy, deeply cut, palmate leaves are not lacking in beauty. But when the spikes of flowers fill the air with the worst of hay-fever-provoking pollen who can see any beauty in Ragweeds? And then would not "food of the devils" be more appropriate than "ambrosia"? The weeds are of real value, however, for where they hold sway they grow rapidly. A wheat field will often produce, in a few days after the wheat is cut, a perfect stand of Ragweeds that fairly begs the farmer to get out his plow and turn it under before the weeds bloom. The ground is hard just then, so if the farmer does anything about it—and if he is subject to hay fever he will do something about it—he gets out his mowing machine and clips the field. By so doing he loses a lot of nitrogen and humus, for Ragweeds are just as acceptable to soil bacteria that make these two soil essentials as are any of the crops we may grow for green manure.

The best way to fight this weed is to use it by plowing it under, but it is an easy weed to kill with chemicals. A spray of iron sulphate in its early stages will completely destroy a crop of Ragweeds—but why use a spray and spend money for fertilizer?

TECHNICAL DESCRIPTION

Ambrosia (Tourn.) L. Fertile heads 1–3 together, sessile in axils of leaves or bracts, at the base of racemes or spikes of sterile heads. Sterile involucres flattish or top-shaped, of 7–12 united bracts, containing 5–20 staminate flowers, with or without slender chaff intermixed. Anthers almost separate. Fertile involucre (fruit) ellipsoid, obovoid, or top-shaped, closed, pointed, resembling an achene and enclosing a single flower; elongated style-branches protruding. Achenes ovoid.—Coarse homely weeds, with opposite or alternate lobed or dissected leaves, and inconspicuous greenish flowers, in late summer and autumn; ours annuals, except the last. (The Greek and later Latin name of several plants, as well as of the food of the gods.)

Ambrosia artemisiifolia L. Much branched, 0.3–2.5 m. high, hairy or roughish-pubescent; *leaves thin, bipinnatifid,* smoothish above, paler or hoary beneath; *fruit* obovoid or globular, *armed with about 6 short acute teeth or spines.*—Roadsides, etc., very common.—Extremely variable, with finely cut leaves, those of the flowering branches often undivided; rarely the spikes all fertile.

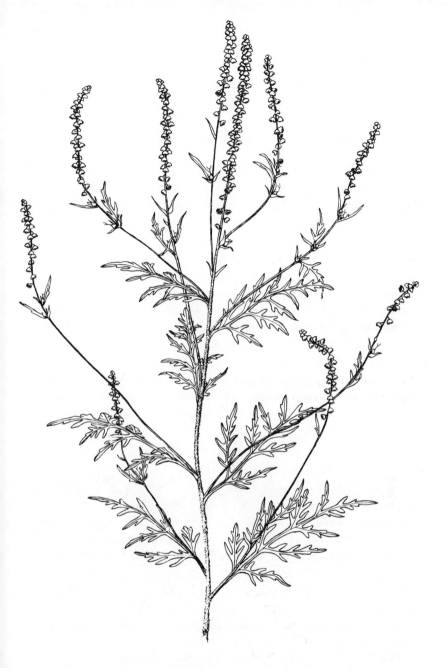

FIG. 91. The Ragweed, the menace of
hay fever patients

COCKLEBUR

[Xanthium orientale L.]

THE cocklebur likes rich soil. It is at its best in rich bottom-lands where overflows scatter the heavy seeds from infested places farther upstream. The farmer who has such land can well despair of ever ridding it of cockleburs, for even if flood waters do not come every year, every bur has in it two seeds, one that grows the first year after ripening and another that germinates a year later. This device means a two-year fight even in upland fields, if a ripened bur ever finds lodgment there.

And they are very likely to find lodgment in upland fields. The hooks on the bur of the cocklebur are a clever device for seed distribution. Every animal that walks through a patch of cockleburs or even brushes against a single plant when its seeds are ripe gives a free ride to as many burry passengers as come in contact with the hair or clothing of that unsuspecting animal. Every hunter and his dog help this culprit over the fences into other fields. The longer the ride the better, so long as there is a chance of reaching the rich corn land of some lazy farmer. For the cocklebur likes cornfields. It likes to be cultivated. If it can get near a big hill of corn it will grow very little until after the corn is laid by. It thus escapes notice. But after the farmer leaves the field it takes up the business of growing with all its weedy vigor, and just before frost comes it fills most of the space, on either side of the corn row that sheltered it, with its heavy branches loaded with burs. The wise farmer will go through his field with a hoe long before frost comes and will take out all such soil sappers, but the lazy farmer, the friend of the cockleburs, lets them grow, and a single plant will thus be able to fill his soil with two years of worry.

The bur of the cocklebur is so different from that of any other plant that the weed may be identified by the bur alone. It can also be identified by its leaves. They are not heart-shaped but nearly so, and are large and thin in comparison with most other

Fig. 92. The Cocklebur ready for frost

weed leaves that are likely to be found growing where the cocklebur grows. The green of the cocklebur leaf is not so green as that of other weed leaves. It has a yellow-green cast.

The cocklebur is a bad weed, but it is easy to identify and easy to eradicate. It is a good soil builder when used as a green manure crop before any of its seeds have a chance to ripen.

TECHNICAL DESCRIPTION

Xanthium (Tourn.) L. Sterile and fertile flowers in different heads, the latter clustered below, the former in short spikes or racemes above. Sterile involucres and flowers as in *Ambrosia,* but the bracts separate and receptacle cylindrical. Fertile involucre coriaceous, ovoid or ellipsoid, clothed with hooked prickles so as to form a rough bur, 2-celled, 2-flowered; the flower consisting of a pistil and slender thread-form corolla. Achenes oblong, flat.—Coarse annuals, with branching stems, and alternate toothed or lobed petioled leaves; flowering in summer and autumn. (Greek name of some plant used to dye the hair; from Greek word meaning yellow.)

Xanthium canadense Mill. Leaves broadly ovate, cordate, usually 3-lobed and simply or doubly dentate; *burs glabrous or merely granular —or glandular-puberulent; the body fusiform-ellipsoid,* 14-17 mm. long, 5-8 mm. in diameter; the *beaks* usually 2, *straight* or but slightly curved; *prickles scattered,* straight-tipped or hooked.—Rich soil, especially in moist places.

WILD SUNFLOWER

[*Helianthus grosseserratus* Martens.]

EVERYBODY knows the common sunflower, *Helianthus annuus;* many people know the Prairie or Kansas sunflower, *Helianthus petiolaris;* but when it comes to the other twenty-two species, unless it be the Jerusalem artichoke, *Helianthus tuberosus,* almost nobody but the botanist has a name for them. The species here treated is a fair illustration. This plant is from seven to twelve feet tall and is seen in neglected fence rows and along highways from Maine to New Jersey, west to the Dakotas and south to Texas, but it has no common name. It should be called the Smooth tal sunflower, the Tall wayside sunflower, the Tall glau-

Fig. 93. The Wild Sunflower is one of our tallest weeds

cous sunflower, or the Tall little-headed sunflower, or anything else that might be descriptive, but it is not so much as called sunflower. If it is given a name at all it is when some one asks, "What is that tall, yellow-flowered weed?"

Well, it is *Helianthus grosseserratus*. It is one of the perennial sunflowers that grows a stalk slender enough, tough enough, and long enough to make a good lance shaft. It is a fine specimen of a weed even if it does not have a very weedy nature. It is too big and too slow to get into cultivated fields, and like most sunflowers it is too much relished by grazing stock to succeed in pastures. If it were not so tall and top-heavy it might be used as an ornamental plant. As it is, however, it is just a wayside weed or a weed of waste places. A great many of its relatives use the same habitats that it does, so we cannot call it the wayside sunflower, and until the common people become weed conscious enough to give it a *viva-voce* name it will have to be called *Helianthus grosseserratus*.

Helianthus is from two Greek words, *helios* meaning sun, and *anthos* meaning flower. *Grosseserratus* is from two Latin words meaning large and rough. The large, rough sunflower would be a good name for it if the plant was always rough. It is rough sometimes with saw-toothed leaves. A close relative is rough, but the name is intended for the species that is smooth.

TECHNICAL DESCRIPTION

Helianthus L. Heads many-flowered; rays several or many, neutral. Involucre imbricated, herbaceous or foliaceous. Receptacle flat or convex; the persistent chaff embracing the 4-sided and laterally compressed smooth achenes, which are neither winged nor margined. Pappus very deciduous, of 2 thin chaffy scales on the principal angles, and sometimes 2 or more small intermediate scales.—Coarse and stout herbs, with solitary or corymbed heads, and yellow rays; flowering toward autumn. (Named from two Greek words meaning *the sun* and *a flower*.)

Helianthus grosseserratus (Martens.) Stem 2–3 m. high; *leaves elongated-lanceolate* or ovate-lanceolate, taper-pointed, sharply serrate or denticulate, acute or attenuate at base, *petioled,* often whiter and finely pubescent beneath; bracts lance-awl-shaped, slightly ciliate.— Dry plains, western Maine to New Jersey, westward to Ontario, Dakota, and Texas.

SPANISH NEEDLES

[*Bidens aristosa* (Michx.) Britton; *Bidens bipinnata* L.]

"O WONDROUS days of green and gold,
A poem here we read."

The poet must have been looking at fields covered with Spanish needles. There are other flowers just as golden but none of them spreads the cloth of gold that some of the Spanish-needle species are able to spread.

No attempt is made here to give the different species of these weeds. *Gray's Manual of Botany* lists fifteen species and four varieties. Most of these are in the Middle Western States, but enough of them are scattered throughout the country to make it possible for every one who has picnicked in the wide-open spaces during October's bright blue weather to get his clothes filled with their sharp-pronged fruits. He might have called them stick-tights, beggar ticks, or beggar lice, but they were the fruit of the beautiful Spanish needle flower masses he admired so much a month earlier.

There are two patterns to these barbed fruits (seeds, in common parlance) of Spanish needles. The bootjack type is the one most often met with. It is nearly always the product of the showy-flowered species. One of the common naked-head types has seeds with four prongs; a sort of Devil's pitchfork arrangement. It is an innocent-looking little weed with leaves so much like those of a marigold or some other cultivated plant that even good gardeners permit it to grow until the flower head appears; then there is no mistaking the culprit. The naked-headed one shown in the drawing is *Bidens bipinnata,* the twice-pinnafied-leaved Spanish needle.

Of course the reason for these pronged fruits is that the seeds may be distributed, and how effective the distribution is one can easily conclude when he sees areas of the yellow blossoms, a great sheet of gold that would put Wordsworth's Daffodils to shame. These are all from the seed of the year before. It required one seed to make every one of those stalks, and every person and every animal that crossed that "bloomin'" field the year before had a part in distributing the seed for the show of this year.

The weeds are worthless—worse than worthless—for they take a prominent part in aggravating the hay-fever cases of their season. They can and should be used as soil builders, however. The farmer who fails to feed his Spanish needles to his hungry soil bacteria deserves to go hungry himself, and he probably will; at least he will miss a great opportunity to raise the fertility of his soil if he fails to plow under his weeds before they bloom.

The name, *bidens,* is Latin for two-toothed. Most of the species do have two teeth on their fruits. The specific name of the naked-headed species given here has already been explained. *Aristosa,* the name of the rayed type here given, and probably the rayed species most often seen, means bristlelike awns, referring to the slender awns of the seed, of course.

TECHNICAL DESCRIPTION

Bidens L. Heads many-flowered; the rays when present 3–8, neutral. Involucre double, the outer commonly large and foliaceous. Receptacle flattish; chaff deciduous with the fruit. Achenes flattened parallel to the bracts of the involucre, or slender and 4-sided (rarely terete), crowned with awns or short teeth (these rarely naked).—Annual or perennial herbs, with opposite various leaves, and mostly yellow flowers. (Latin, *bidens,* two-toothed.)

Bidens aristosa (Michx.) Britton. Somewhat pubescent; leaves 1–2-pinnately 5–7–divided, petioled; leaflets lanceolate, cut-toothed or pinnatifid; heads panicled-corymbose; *outer bracts 8–10, not exceeding the inner, barely ciliate;* rays showy; *achenes with 2* (rarely 4) *long and slender diverging awns* as long as the achene itself or reduced to short teeth.—Swamps, Ohio to Michigan, Minnesota, and southwestward; adventive in waste places eastward. August–October.

Bidens bipinnata L. Smooth annual, branched; leaves 1–3–*pinnately parted,* petioled; leaflets ovate-lanceolate, mostly wedge-shaped at the base; heads small, on slender peduncles; *outer involucre of linear bracts equalling the short pale yellow rays; achenes 4–grooved,* nearly smooth, 3–4–awned, *very unequal.*—Damp soil. Rhode Island, westward and southward; occasional on ballast northward.

Fɪɢ. 94. Two of the Spanish Needles

YARROW

[*Achillia millefolium* L.]

ONE of the commonest weeds of meadow and pasture lands is the yarrow. It is the lacy-leafed, rather fragrant-smelling plant that bears a corymb of small white, sometimes pink, flowers at the top of a lacy-leafed stem. The leaves are so lacy and fernlike that he who sees such a plant either with white or pink flowers, or even with no flowers at all, can be sure he is seeing the yarrow.

The weed has several common names: Milfoil, Thousandleaf, Bloodwort, Old-man's-pepper, Soldier's woundwort, Knight's Milfoil, Thousand weed, Nosebleed, Devil's nettle, Devil's plaything, Badman's plaything, and Yarroway. Most of these are English, and the word Yarrow, which seems to have no other meaning than the designation of this plant, comes down from the Anglo-Saxon. Bloodwort, Nose bleed and Woundwort refer to the fact that the astringent action of the crushed leaves will stop the flow of blood. Yarrow, then, is medicinal in nature. Its tea is used as a tonic, and is said to be a good remedy for severe colds, since it acts as a stimulant and produces perspiration.

The botanical name, *Achillia millefolium* L. means the many-leafed plant, or the thousand-leaf plant, that Achilles used. Perhaps that was why Achilles was invulnerable. He knew how to stanch the blood until an artery was severed. At any rate we know (the dictionary tells us) that he used this weed, or one of its relatives, on the wounds of Telephus.

Yarrow is a composite, but unlike the dandelion and the sunflower its heads of flowers are very small, and they are grouped in flat-topped corymbs ("A corymb is a flat-topped, or convex, open, flower cluster") that the layman always takes to be the flower. The dandelion and sunflower both have many small flowers in a big head. The Yarrow has *few* small flowers in *many little* heads. When one discovers these facts he also discovers that each head has, around and below the flowers it contains, a structure similar to a calyx, except that it is made up differently. This

Fig. 95. The Yarrow, beautiful and worthless

is what is called an involucre, and it is the distinguishing mark of all composites. The bracts (parts) of the Yarrow involucre are hairy.

This plant is about as worthless as any that grows. It is not eaten by stock, and since it is never in cultivated fields it cannot be used as a fertilizer. The little medicinal value the weed has is not worth the place it occupies. The pink-flowered variety is sometimes used as a decorative plant, however, and so Yarrow might be classified as a weedy aristocrat, but as a very weedy one.

TECHNICAL DESCRIPTION

Achillea L. Heads many-flowered, radiate; the rays few, fertile. Involucral bracts imbricated, with scarious margins. Receptacle chaffy, flattish. Achenes oblong, flattened, margined; pappus none.—Perennial herbs, with small corymbose heads. (So named because its virtues are said to have been discovered by Achilles.)

Achillea millefolium L. Stem simple or sometimes forked above, 3–10 dm. high, *arachnoid or nearly smooth; stem-leaves numerous* (8–15) smooth or loosely pubescent; *corymbs* very compound, 6–20 cm. broad, *flat-topped,* the branches stiff; involucre 3–5 mm. long, its bracts all pale, or in exposed situations the uppermost becoming dark-margined; rays 5–10, white to crimson, short-oblong, 1.5–2.5 mm. long.— Fields and river-banks, common. (Eurasia.)

DOG FENNEL

[*Anthemis cotula* L.]

DOG FENNEL, like the dandelion and the sunflower, belongs to the great family of composites. A composite is a flower that is made up of flowers. The flower of the dog fennel is in reality a great number of little yellow flowers surrounded by a row of little flowers, each of which bears a single white petal. This is the way the composites have solved their seed problem. There is but one seed to each flower, but there is a great number of flowers and so there is a great number of seeds produced by every "flower head."

FIG. 96. Dog Fennel or Mayweed

The weed is as easily recognized and remembered by its smell as by its appearance. A single dog-fennel plant is not much to look at, but one seldom sees it as a single plant. It is usually in a mass, a veritable mob, assailing some henhouse or pigsty, and the air as well, with its unforgettable odor. Its odor is not so sickening as that of the Jimson weed which often accompanies the dog fennel in its attacks, but it is so bad that chickens—animals that seemingly are never bothered by the smell of anything they eat— never touch dog fennel. And fleas, mosquitoes, and bees are repelled by it. Extract of the weed is classified as a vegetable poison, and the crushed leaves are said to be liable to blister. The tea brings sleep to asthma patients, and copious draughts of it cause sweating and vomiting. The weed is, therefore, medicinal.

The lacy leaves, made so by being "finely 3-pinnately dissected" (as we find in the technical description), the yellow-centered, white-margined flower heads, and the never-to-be-forgotten odor of the bruised leaves are enough to identify the plant. It is seldom much over a foot in height, and it is nearly always found in abundance when it is found at all, crowding out all other herbaceous vegetation. Clean mowing before the flowers open is all that is required to eradicate the weed.

Dog fennel has several common names, as have nearly all weeds that came to us by the way of England. It is called Wild chamomile, Dog's chamomile, Fetid or Stinking chamomile, Pigstyweed, Mayweed, Stinking daisy, Hog's fennel, and Chamomile. Chamomile seems to refer to the smell of the true chamomile flowers, which scent is supposed to suggest that of the earth-apple—whatever that may be. The generic name, *Anthemis,* is the ancient Greek name for the Chamomile. *Cotula,* the name of this species, is a Roman name for a peculiar kind of cup, but how it applies to this plant is difficult to see.

TECHNICAL DESCRIPTION

Anthemis L. Heads many-flowered, radiate; rays pistillate or neutral. Involucre hemispherical, of many small imbricated dry and scarious bracts shorter than the disk. Receptacle conical, usually with slender chaff at least near the summit. Achenes terete or ribbed, glabrous, truncate; pappus none or a minute crown.—Branching often strong-scented herbs, with pinnately dissected leaves and solitary terminal heads; rays white or yellow (rarely wanting); disk yellow. (The ancient Greek name of the Chamomile.)

Anthemis cotula L. Annual, acrid, *ill-scented;* leaves finely 3-pinnately dissected; rays mostly neutral; *receptacle without chaff near the margin;* pappus none; *achenes tuberculate-roughened.* Common by roadsides. (Naturalized from Europe.)

OX-EYE DAISY

[*Chrysanthemum leucanthemum* L.]

THE Ox-eye daisy is a beautiful, bad weed. It can and does adorn many a flower garden, but when it climbs over the fence and ambitiously tries to adorn a whole meadow, that is just too much adornment. For after all a meadow is not a flower garden, and beauty is as beauty does even among plants. A cow is not sentimental, and daisies in the hay are just that much less hay to her; to her keeper they are just that much less milk.

The pest is a near relative of all the chrysanthemums that elicit our superlative adoration at "mum" shows, and some of the varieties of this particular species are but a little lower than those angelic hosts. It is the weedy variety that we are considering, however, and it has been a weedy variety for a long time. If we care to know what some of our English progenitors thought of the Ox-eye daisy we have only to glance at most of the names they gave it. Here are a few of them: White weed, Dog daisy, Bull daisy, Poorland daisy, Maudlin daisy, Butter daisy (meaning, it spoiled the butter), Poverty weed, Dog blow, and Moon penny.

It is true that there were those who saw only the beauty of the plant; perhaps old-maid school teachers—both male and female— and so we find it called Field daisy and Marguerite. Carolus Linnæus, the great namer of plants and animals, called the weed *Chrysanthemum leucanthemum,* which, translated, means the white, gold flower, or the white-flowered gold flower. Chrysanthemum is from two Greek words, modified by the Latin and French, *Chrysos*, gold, and *anthemon, flower. Leucanthemum* is also made from two Greek words, *leuc* meaning white, and the *anthemon* again meaning flower.

As we might expect of a weed so conspicuous, the Ox-eye daisy was early found to have medicinal properties. The whole plant and the flowers are used in both the tea and extract form—if there is an extract. The tea is said to act as an antispasmodic, as a diuretic, and as a tonic. It was, and perhaps still is, used as a medicine for whooping cough, and for cases of asthma and nervous excitation as well. The weed can be made into lotions and used externally for wounds and bruises. There seems to be no market for Ox-eye daisy, and so we may conclude that it is its use principally, if not entirely, as a source of home remedies that accounts for its medicinal reputation.

Daisies are so well known to those who have any interest in flowers that a description of the Ox-eye daisy would be superfluous. Suffice it to say that any daisylike flower in abundance in any meadow or neglected park is almost sure to belong to the Ox-eye. Even the leaves are so characteristic that a look at the drawing will make the identification of the plant almost, if not quite, certain. The flowers (actually flower heads) are from one and one-half inches to two inches in diameter, and are single at the ends of rather long, slender stems.

The best method of fighting such perennials as the Ox-eye daisy is that of plowing up the meadow or pasture land just before the flowers are set. Keep the field in cultivated crops for three or four years and then reseed to grass.

TECHNICAL DESCRIPTION

Chrysanthemum (Tourn.) L. Heads many-flowered; rays numerous, fertile. Scales of the broad and flat involucre imbricated, with scarious margins. Receptacle flat or convex, naked. Disk-corollas with a flattened tube. Achenes of disk and ray similar, striate.—Annual or perennial herbs, with toothed, pinnatifid, or divided leaves, and single or

FIG. 97. The Ox-eye Daisy is a chrysanthemum

corymbed heads. Rays white or yellow (rarely wanting); disk yellow. (Old Greek name meaning golden flower.)

Chrysanthemum leucanthemum L. Stem erect, simple or forked toward the summit; basal leaves spatulate-obovate, on long slender petioles, the blades crenate-dentate; middle and upper stem-leaves oblong or oblanceolate, coarsely and regularly crenate or dentate above, with larger spreading teeth at base; heads 4–6 cm. broad; *involucral bracts narrow, brown-margined; rays white* (rarely tubular, laciniate, or deformed). Fields, etc., Newfoundland and eastern Quebec to New Jersey; rare southward. June–August. (Naturalized from Europe.)

BURDOCK

[*Arctium lappa* L.]

BURDOCK is one of the big weeds. A single stalk may attain a height of from six to eight feet and be so branched and heavy with leaves and burs that it will occupy a space of from six to eight feet in diameter. The plant is a biennial and in the first year stores in a long root all the food that its many broad leaves can make. These leaves are from six to eight inches broad and from fifteen to eighteen inches long, and they are spread so as to get all the sunlight and carbon dioxide it is possible to get for the sugar manufacturing that goes on in them. After a year of such preparation it is not strange that a treelike stalk can be sent up from that sugar-filled root. The crop of seed developed in the second year, however, not only exhausts the food supply of the root, but also all the food that the foliage of the stalk can produce—no small amount—and so the burdock dies from exhaustion after the maturing of the seed crop which every hairy animal that brushes against it will carry to a new location. There is always a plentiful supply left at the old location, with the result that a clump of burdock never grows any smaller, unless man takes a hand in the matter.

It is its food-filled root that puts the burdock in Pharmacopœia. The roots are raised for the market in Europe, and the United States imports from twenty-five to fifty tons of dried lappa (bur-

FIG. 98. A Burdock sprig in front of
one of its big leaves

dock roots) each year. Lappa is used in blood medicines, and is said to have effected a cure in many cases of eczema—an almost incurable disease. It is also used as a diuretic and an alterative.

The common names of the weed are nearly all descriptive: Fox's clote, Beggar's buttons, Cockle-buttons, Stick-buttons, Hardock, and Bardane, but to him who knows that "dock" is usually applied to a broad-leafed plant, "Burdock" is just as descriptive as any of the other names. The *Arctium* of the botanical name is from the Greek word *arktos,* which means a bear. *Lappa,* the specific name, is from the Latin and means a bur. *Arctium lappa* L. then means the burry bear named by Linnæus. It is a very good name for the plant when one knows the burs. Rural children make chains and baskets of the blooming burs, and sometimes one of them is mean enough to throw a bunch of the burs into the hair of a rival, or even into the hair of the girl he thinks has snubbed him. She who has had this experience needs no technical description of the Burdock.

The root of the weed is used in medicine; its stalk is said to be used as a food in some parts of the world. When the rind is stripped off and the internal part of the stalk is cooked it tastes like asparagus, but it is highly laxative and so very limited amounts may be eaten by most people. This same core may be candied, and is said to be nutritious and wholesome.

TECHNICAL DESCRIPTION

Arctium L. Heads many-flowered; flowers all tubular, perfect, similar. Involucre globular; the imbricated bracts coriaceous and appressed at base, attenuate to long stiff points with hooked tips. Receptacle bristly. Achenes oblong, flattened, wrinkled transversely; pappus short, of numerous rough bristles, separate and deciduous.—Coarse biennial weeds, with large unarmed petioled, roundish or ovate, mostly cordate leaves, floccose-tomentose beneath, and small solitary or clustered heads; flowers purple, rarely white. (Name from the Greek meaning *a bear,* from the rough involucre.)

Arctium lappa L. *Head subcorymbose, 3-5 cm. broad;* involucre glabrous; *bracts straightish,* lance-to-linear-attenuate. Roadsides and waste places, widespread. (Introduced from Europe.)

CANADA THISTLE

[Cirsium arvense Scop.]

THERE is no weed worse than the Canada thistle, especially in our northern States and Canada. Fortunately it does not thrive very far south. It is a sort of Eskimo in its bristly coat, and so shrinks from the hot sun. Or is it because its photosynthetic processes require the long days of the North? Whatever it is that keeps it in its favorite latitudes, they so completely agree with the plant that in all sorts of habitats there it is the very worst weed. And it is, perhaps, the worst weed of the entire United States. The plant does not have a single virtue so far as man is concerned. Its seeds serve as bird food for goldfinches and sparrows, but even this good turn is offset by the fact that the birds in getting their food set free the winged seeds and wherever those seeds fall trouble begins.

The plant is outlawed in every northern State; thirty-seven States in all legislate against this rogue, but outlawing it has had very little effect upon it. It takes pastures and hayfields, and even enters and stays in cultivated fields in spite of the fact that it is a perennial. It can be eradicated if hoed off every week for a year or two, and some have succeeded in eradicating patches of it by smothering them with strawstacks and with mulching paper, but no simple method of eradication works with the Canada thistle.

The thing that makes the weed so pernicious is its creeping rootstalks. A single seed, because of these, can develop a colony of unlimited size if unmolested for a few years. Fortunately the plant is dioecious; that is, there are two kinds of plants, male and female, and unless both kinds are together viable seeds are not produced. If a colony is started by a seed that makes a male plant, then every weed in the colony will bear male flowers and no seed will be produced by that patch regardless of its size. If the seed produces a female plant, and there are no male plants near, then no matter how many plants are in that colony of females no fertile seeds will be developed by them. This helps to explain the fact that many of the patches of Canada thistle in the States south of the weed's normal range lack viable seeds.

But the Canada thistle does not need seeds. A small bit of a rootstalk can start a colony quite as well as a seed can, and plows and cultivators make glad this rogue of rogues among weeds.

So many people expect to see a big, bristling plant when they look for Canada thistles, and so many disregard the patches of short thistlelike plants they may discover in a lot or field, that a description of the weed is necessary. The drawing accompanying this sketch does not show the flower heads. One should know the plant without its flowers. It blooms late in the season, long after it has done most of its spreading. The weed should be recognized early, when it appears as the picture shows it, in shoots of from four inches to eighteen inches high. When a colony of little spiny-edged-leaved plants appears in a lot or field it is the Canada thistle. No other thistlelike weed is found in patches of little and big, tall and short plants. Of course the thistlelike flowers will appear later, but not until dozens of shoots have popped up farther and farther away from the center of the colony. They arise from those rootstalks that are burrowing like moles under the ground. If one has any doubts as to the identity of the plant he has only to dig or pull up a few of the weeds. If there are two or more stalks on what he might call a long side root, and if those stalks have prickly-edged leaves, he can rest assured he has found the culprit. Later he may know it by its thistlelike heads of flowers that are smaller and more numerous than are those of the common or pasture thistle, and the plants themselves are smaller. The Canada thistle averages about two feet in height when in bloom; the common thistle has often a height of from three to four feet.

The weed has several descriptive names: the Small-flowered thistle is one of the best; Creeping thistle is another good one, and it is also called Green thistle and Perennial thistle. Its botanical name, *Cirsium arvense,* means the field plant that will cure swelling veins. *Cirsium* is Greek for "a swelling vein" and *arvense* refers to fields, especially cultivated fields.

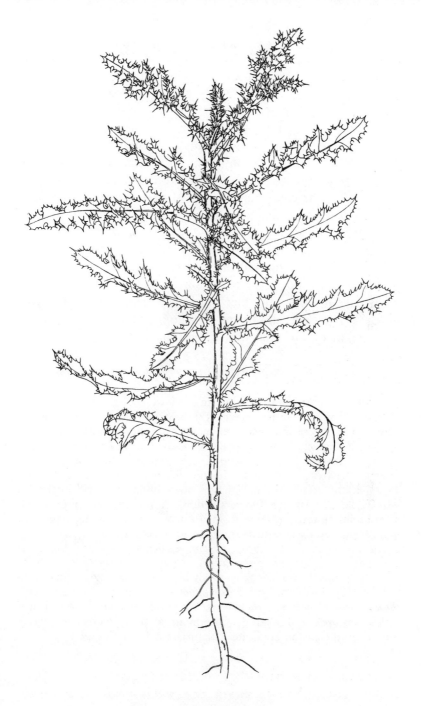

Fig. 99. A young stalk of the
infamous Canada Thistle

Something has already been said concerning the control of the weed. This is one weed on which any and all control methods can and should be used. Small patches can be destroyed by hoeing off the shoots as fast as they appear, until they appear no more. Sodium chlorate may be used either as a spray or as a salt; as a spray on the young shoot in the spring and summer, and as salt on the ground in late autumn. The weed can be controlled by cultivation if the cultivating is done often and carefully enough. Infested pastures should be plowed up and if possible used as cultivated crop lands for two or three years, when, if the last thistle is destroyed, they may be turned back to pasture.

It should be remembered that any plant can be killed—starved to death—if it is not permitted to spread its leaves for more than a few days at a time. In other words, persistent cutting of weeds, once every week let us say, for a single season will starve out the worst of them, and that means the Canada thistle.

TECHNICAL DESCRIPTION

Cirsium (Tourn.) Hill. Heads many-flowered; flowers all tubular, perfect and similar, rarely imperfectly diœcious. Bracts of the ovoid or spherical involucre imbricated in many rows, tipped with a point or prickle. Receptacle thickly clothed with soft bristles or hairs. Achenes oblong, flattish, not ribbed; pappus of numerous bristles united into a ring at the base, plumose to the middle, deciduous.—Herbs, mostly biennial; the sessile alternate leaves often pinnatifid, prickly. Heads usually large, terminal. Flowers reddish-purple, rarely white or yellowish; in summer. (Name from a Greek word meaning *a swelled vein,* for which the Thistle was a reputed remedy.)

Cirsium arvense (L.) Scop. Perennial, slender, 3–9 dm. high, the *rootstalk extensively creeping; leaves oblong or lanceolate,* smooth, or slightly woolly beneath, finally green both sides, strongly sinuate-pinnatifid, very prickly margined, *the upper sessile but scarcely decurrent; heads imperfectly diœcious;* flowers rose-purple or whitish.—Cultivated fields, pastures, and roadsides, common; a most troublesome weed, extremely difficult to eradicate. (Naturalized from Europe.)

CHICORY

[Cichorium intybus L.]

CHICORY, Succory, Blue sailors, Blue daisy, Coffee weed, and Bunk are some of the common names of a common roadside and meadow weed. It is found in nearly all parts of the United States and Canada, but it is partial to limestone soil and so may not be found in every locality. It is a perennial and for this reason does not molest cultivated fields and gardens. In fact it can scarcely be said to molest anything, for if it gets into a meadow—the only place it might cause trouble—it makes good hay. The plant comes to us from Europe, where it is considered no more of a weed than sweet clover is in this country. It is used as a hay crop there, and the farmers who grow it declare that it is superior to alfalfa; that they are able to cut more hay from a chicory field in four cuttings than they can take in the same number of cuttings from the best field of the queen of legumes.

Chicory is also cultivated for its roots and early shoots. The roots are dried and used as a substitute for and as an adulterant of coffee. They are used as a vegetable, too. The shoots are forced and blanched for salads. Some very fine varieties of chicory have been selected for root culture and shoot forcing. The varieties Magdeburg, Brunswick, and Zealand have big roots, while the Witloof is a variety with tasty, tender leaves. Chicory and endive are brothers—or sisters—and so it is not strange that chicory is an edible plant.

The weed is easily recognized by its blue, composite flowers that spread themselves in the morning along the sparsely leaved stems. The color of the flowers is so strikingly blue that the flower heads alone should identify the plant, but it has another character that with the flowers makes identification certain. The plant has a milky juice. Blue flower heads as big as silver dollars strung along sparsely leafed stems that are filled with a milky juice, that is chicory and nothing else.

The generic name, *Cichorium* (pronounced Sĭ-kō'rĭ-ŭm), is,

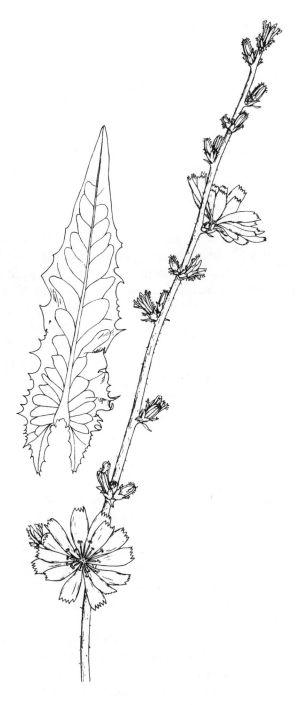

FIG. 100. Chicory or Blue Sailors

according to one authority, from the Arabian; by another it is said
to be Egyptian. The specific name, *intybus,* is the modification
of another eastern name, *Hendibeh,* which refers not only to this
plant but to endive as well. In other words the scientific name of
this weed is so lacking in descriptive significance that he who
never tries to say *"Cichorium intybus* L." will find others quite
as lacking in erudition as he.

TECHNICAL DESCRIPTION

Cichorium (Tourn.) L. Heads several-flowered. Involucre double,
herbaceous, the inner of 8-10, the outer of 5 short and spreading bracts.
Achenes striate; pappus of numerous small chaffy scales, forming a short
crown.—Branching perennials, with deep roots; the sessile heads 2 or 3
together, axillary and terminal, or solitary on short thickened branches.
Flowers bright blue, varying to purple or pink (rarely white), showy.
(Altered from the Arabian name of the plant.)
Cichorium intybus L. Stem-leaves oblong or lanceolate, partly clasp-
ing, the lowest runicate, those of the rigid flowering branches minute.
—Roadsides and fields, Newfoundland to Minnesota, and southward.
July–October. (Naturalized from Europe.)

DANDELION

[*Taraxacum officinale* Weber.]

DANDELION, Blow ball, Lion's tooth, Peasant's cloak, Yellow gowan,
Priest's crown, Irish daisy, and Monk's head are some of the
names given this common lawn weed.

Just why the plant is called dandelion (tooth of the lion) may
be hard to decipher without considerable imagination. The name
refers to the toothed edges of the leaves. Some of these, when
attention is called to them, certainly do resemble the fangs of
that noble cat.

The botanical name of the plant is much harder to explain.
Taraxacum seems to be of Persian origin, and probably refers to
some plant found in Persia, or perhaps to the medicinal properties
of some plant. *Officinale,* the name of the larger species, does

refer to the medicinal properties of the dandelion. Whenever this name, *officinale,* is given to any species of plant it means that this particular species is used by druggists or pharmacists. The other species of dandelion, called *Taraxacum erythrospermum* by most botanists, means the red-seeded Taraxacum, or red-seeded dandelion.

Nearly every one knows the dandelion because of its beautiful blooms, those blooms that are the special irritant of lawn makers. For a lawn may be looking as free from dandelions as a Brussels carpet after the mowing, and the next morning be covered with golden-headed stipes four or five inches tall. A day later the blow balls will be blowing, and the weed will have completed its weedy work. For this it was made: to grow near the ground where all the water and carbon dioxide used in food making can be easily obtained, to make all the food it can and to stuff it all in its fat root, and then to use this stored food, or all of the energy that may be released from it, to send up its seed scapes before the angry lawn maker thinks of cutting his grass again.

But the dandelion is not entirely bad, and the truth is the lawn maker pays little attention to it except when it blooms in the early spring, and again after the fall rains. Its leaves pass unnoticed in the grass throughout the summer, and if the grass sod is what it should be the dandelions are not able to establish themselves there. They thrive only where the grass is sparse and where space enough is left for the weeds to get in. In weak lawns they soon become a nuisance, however, and when they become too numerous to be disregarded the best way to rid the lawn of them is to plow it up, work it down and reseed it. It will take a whole year to make a weedless lawn in this way, but sometimes this is the only solution of the weed problem.

The dandelion is not so lacking in virtues as are its associates, the plantains. Its flowers have beauty and its roots have worth. More than a hundred thousand pounds of dandelion roots are imported by the United States each year. They are used by pharmacists in tonics and in liver medicines. The flowers of the plant are used for making dandelion wine, and the leaves are used for greens. The dandelion is a weed, but a virtuous one.

FIG. 101. The Dandelion in its weediest way

TECHNICAL DESCRIPTION

There are two kinds of dandelions, *Taraxacum officinale* and *Taraxacum erythrospermum*, but they are so nearly alike that a technical description of one of them is sufficient. The general description of *Taraxacum* is as follows:

Taraxacum (Haller) Ludwig. Heads many-flowered, large, solitary on a slender hollow scape. Involucre double, the outer of short bracts; the inner of long linear bracts, erect in a single row. Achenes oblong-ovate to fusiform, 4–5–ribbed, the ribs roughened; the apex prolonged into a very slender beak, bearing the copious soft white capillary pappus. —Perennials or biennials; leaves radical, pinnatifid or runcinate; flowers yellow. (Name from the Greek word meaning *to disquiet* or *disorder*, in allusion to medicinal properties.)

Taraxacum officinale Weber. *Leaves coarsely pinnatifid*, sinuate-dentate, rarely subentire; *heads large* (3–5 cm. *broad*), *orange-yellow;* involucral *bracts not glaucous; the outer elongated, conspicuously reflexed; achene olive-green or brownish*, bluntly muricate above, its beak 2–3 times its length; *pappus white.*—Pastures and fields, very common. April–September (and rarely throughout autumn and winter).—After blossoming, the inner involucre closes, and the slender beak elongates and raises up the pappus while the fruit is forming; the whole involucre is then reflexed, exposing to the wind the naked fruits, with the pappus in an open globular head. (Naturalized from Europe.)

WILD LETTUCE

[*Lactuca scariola* L.]

WILD lettuce is the weed that has learned to use both sides of its leaves in its work of food making. The leaves are turned up on edge and take northerly and southerly directions. That is why it is called the compass plant. If a lost man could decide which was north and which was south when he looked at the weed he would have a good compass in the Wild lettuce, and the plant would deserve the name of compass plant; but such is not the case. The leaves pointing north look exactly like those pointing south, and so the plant is just as weedy as a guide as it is in some of its other characters.

Supplied with an abundance of winged seeds that germinate in any and all places, and equipped with the ability to grow rapidly (those double-working leaves take care of that) the Wild lettuce can be expected to show up anywhere. It is often seen elbowing its way among even stronger weeds than it is. It delights in half-tended cornfields and gardens, and it is at its best in fence rows and along waysides. It may attain a height of six feet, but in un-hindered places it prefers to spend its energy in producing branches and flowers, and so its usual height is from three to four feet.

The leaves are prickly, especially along the midribs. The flower heads (this is a composite and so the flowers are really flower heads) are small with pale yellow rays. The leaves and stems are filled with a milky juice that is bitter to the taste, and the stem, especially near the bottom, is rough with prickles. These three characters: small flower heads with pale yellow rays, leaves and stems filled with a bitter milky juice, and prickly leaves and stems, identify this very common weed.

Wild lettuce has some virtues as well as faults. Lettuce opium, lactucarium, is extracted from this very plant. Lactucarium was at one time in the *U. S. Pharmacopœia.* It was and still may be used as an anodyne, a diophoretic, a laxative, and as a diuretic. All of these medicinal properties are in the leaves when eaten as greens (pot herbs) or as salad, but unless one has the appetite of a horse he can scarcely expect to assimilate enough lettuce opium from a dish of greens to obtain all of the curative effects here listed.

Several names have been given to this particular species of wild lettuce. The reader must know that the lettuce genus is a big one and that several of its species are called "Wild lettuce." This species is the most common and most widely distributed, however, and since it comes to us from Europe we might expect it to have several common names. It is called Prickly lettuce, Compass plant, Horse thistle, Milk thistle, and Wild opium. The generic name, *Lactuca,* is the ancient name for lettuce and means

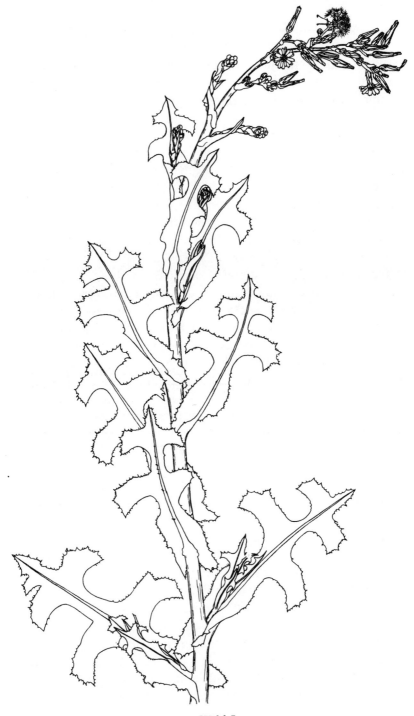

FIG. 102. Wild Lettuce

milk-giving. *Scariola*, the specific name, is evidently of Latin origin, but it has long been used by the English. According to the Oxford Dictionary, as far back as the year 1400 some Englishman said, "Wylde letus hat feldman clepyn Skariole." Wild lettuce have field-men (farmers) called Scariole, is what he said. *Scariola*, therefore, was a widely used name, and perhaps it was for this reason that Linnæus decided upon it for the specific name of this widely distributed plant.

TECHNICAL DESCRIPTION

Lactuca (Tourn.) L. Heads several-many-flowered. Involucre cylindrical or in fruit conical; bracts imbricated in 2 or more sets of unequal lengths. Achenes contracted into a beak, which is dilated at the apex, bearing a copious and fugacious very soft capillary pappus, its bristles falling separately.—Leafy-stemmed herbs, with panicled heads; flowers of variable color, produced in summer and autumn. (The ancient name of the Lettuce, L. *sativa* L.; from *lac*, milk, in allusion to the milky juice.)

Lactuca scariola L. Stem below sparsely prickly-bristly; *leaves pinnatifid, spinulose-denticulate*, tending to turn into a vertical position (*i.e.*, with one edge up); midrib usually setose beneath; panicle loose, with widely spreading branches; flowers pale yellow, sometimes turning bluish in fading or drying.—Roadsides, railway ballast, etc., southern New England to Ohio, Missouri, and Kentucky, chiefly westward, but even there less common than the following variety. (Adventive from Europe.) Variety *Integrata* Gren. and Godr. *Leaves oblong, denticulate*, none of them or only the lowest pinnatifid; midrib pricklysetose or rarely smoothish. (*L. virosa* of Am., auth., not L.)—waste grounds and roadsides, across the continent; westward an abundant and pernicious weed. (Naturalized from Europe.)

IV

WEED CONTROL

WEED CONTROL

How to use weeds, especially certain weed species, has been emphasized in the text: pages 98, 119, 124, 138, etc. But how to prevent them from taking over an area—a lawn, a putting green, a garden, a cultivated field, or even pasture land—without an inordinate amount of labor is something every discerning owner of such property craves to know. Almost every lawn-owner and every caretaker of "greens" knows that there are on the market chemical preparations designated as selective weed-killing sprays, but he often does not know how to use them, nor when to use them. He is told that certain sprays kill only the broad-leaved plants, the weeds that are "not grass-like," and that there are others that kill only the grasses. This means that if the reader has a lawn filled with Dandelions (P. 299) and Chickweeds (P. 110), he can and should kill these weeds with a broadleaf weed spray, and the grasses of the lawn will not be injured thereby. In fact the grasses may be stimulated by this spray.

But the broad-leaf spray will have no effect, unless it be a stimulating one, on the grass-like weeds: Crab grass (P. 29), Nimble Will P. 43), Goose grass (P. 50), etc., all of which find a lawn habitat much to their liking. If they are to be sprayed out—and they can be—with a grass-killing spray, the lawn grasses will also be completely killed, or badly injured.

So the problem of keeping a lawn free from weeds is not an easy one to solve. It can be solved, however, but like all tough problems it takes determination and hard work to accomplish a satisfactory solution.

The home-owner's lawn, without a doubt, posits the most difficult weed problem to the greatest number of people that aggressive weeds have found. But much of the problem can be and should be solved in the making of the lawn. It takes time to make a good lawn, much more time than the average American lawn-owner cares to take, but it is the only way to reduce to a minimum the weed population of a newly made lawn.

The preparation for the lawn should start with the plowing up and working down, and usually with a liming of the soil, of the lawn area at least one full year before the grass seed is sowed or planted. Let us say that the plowing up, liming and working down of the soil takes place in late October or early November. This is the time it should be done in the central areas of the United States. It should be done earlier in northern latitudes, later in the southern. Then, throughout the following year, at least every month from February until October, the lawn area should be harrowed, or raked, and kept moist so that the weed seeds will sprout and the weeds grow high enough to be killed with a spray or sprays.

After the first crop of weeds is killed, the area should be harrowed again to bring other weed seeds up to where they, too, will sprout and grow and be sprayed out. This routine of work—harrowing up seeds and spraying out weeds each following month, oftener if the weeds cooperate—should continue until October. And it must be remembered that the weeds to be sprayed out will be both the grass-like and the non-grass-like: the narrow grass-leaved and the broad-leaved weeds. This means the use of the two types of spray materials: the grass-leaf spray and the broad-leaf spray. It is often said that the two cannot be used together; but greens-keepers declare they have mixed them and obtained good kills, many times quite as good as when they use the chemicals separately. That is, both types of weeds were killed with one spray application when the two types of spray materials were mixed and used as a single spray.

Now comes the seeding of the lawn. After the last spraying out, near the first of October, the soil of the lawn area should be in good shape—in good heart, as the English say—for seeding. The surface of the area should be firm and level, but with contour enough to permit drainage without erosion. If this cannot be, if the lawn area has too much slope, then, of course, terracing or "stripping" must be resorted to. Stripping is done by laying one or more strips of good grass sod on the contour across the lawn area. But this is not a weed problem. It is, however, a grass problem and especially a lawn problem.

The grass seed: Blue grass, Bermuda grass, Perennial rye grass, or whatever may be the species of grass sowed, should contain some seed of an inconspicuous legume; such as White clover, Black Medic clover, or Yellow Hop clover. These three clovers, or any one of them, can be, and should be, used with any one lawn grass or combination of lawn grasses sowed. Lawn seed mixtures usually contain White clover (P. 131). The grasses will make better growth if the clover is there. This is Mother Nature's way. All of her grass lands, prairie grass fields and permanent pasture and hay fields, contain legumes. The legumes are there to supply the nitrogen compounds that all non-leguminous plants must have furnished them.

Of course the lawn-owner can supply this grass need, this nitrogenous need, with commercial fertilizers, spread on the lawn area each year in late winter or early spring. But many lawn-owners like the grass-legume combination, and old Mother Nature likes it. She even puts poisonous legumes and prickly legume species in her prairie grass lands to keep the grazing animals from destroying that available nitrogen supply her grasses must have. So, if you, Lawn Owner, wish to work in harmony with Mother Nature you will use a lawn seed mixture, one containing White clover or some other inconspicuous, perennial legume when you seed your lawn.

Weed control, however, rather than grass culture is our subject; but it so happens that there is a weed control method that ties in with grass culture: with maintaining the nitrogen supply that all grasses require.

Two of the nitrogenous compounds produced by fertilizer companies, ammonium sulphate and ammonium nitrate, are volatile enough—when either of them is spread on the lawn in the spring soon after the grasses and seedling weeds have started—to produce fumes sufficient to kill the young growth. Both the seedling weeds and the lawn grass blades are killed by the fumes, but the perennial grass will soon reappear, fed by the nitrates that have been absorbed by the soil. This is a weed control method every lawn-owner and greens-keeper should know and use. But, it should be used only after the lawn or "green" has been well established, at least two years old, and it should be used only once each year on perennial grasses. The time to use it is in the spring of the year, after the grass and the seedling weeds have reached a height of two or more inches. It is then that the nitrogenous compound is spread, either in the dry state, using a pound or so to each 100 square feet, or sprayed on, dissolved in water.

Of course this method of weed control can be used in per-manent pasture areas, and is often used there; and it can also be used as a pre-emergent weed control with row crops: corn, soy-beans, vegetables, garden crops, etc. "Pre-emergent" in this case refers to the crop plants. The weed crop should be well started when the nitrogenous compound is spread i.e. sprayed on. In other words, the field should be prepared for the crop plants, but not seeded to them, or at least they should not be allowed to appear above the ground before the nitrogenous treatment. The ni-trate may be disced into the soil as soon as the weeds are killed. It then becomes a fertilizer. This nitrate is readily absorbed by the soil, however, and so requires no mechanical mixing with it.

Before closing this chapter with the chemical control of weeds, I should like to make a plea for the use of weeds as a weed control. It has been my good fortune to be compelled to make soil where there was no soil, and where everyone who knew the area involved declared it was quite impossible to grow any-thing, not even weeds, in such places. Of course the natives of these areas (and there were several such areas in seven different states of the United States), were always surprised that a man with my education would attempt such an impossible thing. In every at-tempt, however, within five or at the most six years after starting the project with the plowing of the field, and the spreading there-on from five to eight tons per acre of a high grade, finely pulver-ized limestone, there was a beautiful field of alfalfa to show for our efforts. And what was used to make this soil? Just weeds, principally.

The procedure of developing these soil areas started, as stated above, with the plowing of the field and the spreading of the limestone on the plowed surfaces. The limestone was disced in, and the field was allowed to lie untouched until time for seeding a crop of Winter rye. By that time there would be a few patches of weeds scattered over the field. These were incorporated with the dirt, which was not yet soil, in preparing the field for seeding it to rye.

The rye, a hardy biennial grass, would make a sparse showing over all of the field before winter came, but always a much better showing where the weed patches had been. During the spring months following the seeding of the rye, the rye and weeds, especially the weeds, would make a fairly uniform, green covering of the field, a covering that was plowed under as soon as the scattered rye began to "head."

The field was then prepared for a crop, but the crop was just weeds. The weeds were allowed to grow until time to prepare the field for another crop of rye. They were then plowed under and the field was prepared for, and again sowed to, rye. The time of seeding was usually about mid-September.

The following spring, in early March, well inoculated Red clover seed was sowed in the rye. The clover was to be a test crop. If any clover "stuck," and it usually did "in spots," the field was again treated as it had been the year before, but this time it was sowed to wheat. In the wheat, the following spring, Sweet clover (P. 135), well inoculated seed, was sowed on the field and rolled in, if the field happened to be in condition for rolling.

Usually a surprisingly good wheat crop, containing few weeds and a good "stand" of Sweet clover, indicated that we had already developed a soil.

The wheat was harvested, and the Sweet clover, a biennial, was permitted to grow until it reached the blooming stage the following year. Then it was plowed under and the field was kept "worked down" and free of weeds until the time arrived for seeding the alfalfa. The results obtained by this method of soil building, without the aid of commercial fertilizers, always astonished those who knew the area before the project started. The year following the seeding of the alfalfa at least one good hay crop, and sometimes two crops, were harvested, both free of weeds.

So here is a weed control obtained by using the weeds. But the point I wish to make is the value of weeds in soil building. Weeds plowed into the soil, or hoed into it, are quite as valuable as legumes used in this way. This assertion is likely to shock some readers, but it is true, nevertheless. There is just as much protein material in the cells of a Pigweed or a Lamb's Quarters weed as there is in the same-sized Sweet clover plant. And the protoplasm, the protein material of these decaying and decayed weeds in the soil, is just as available to the nitrate-fixing microflora and microfauna of the soil as are the decaying and decayed protoplasmic contents of the Sweet clover's cells.

This fact was proved several times by soil building projects like the one described above, but most effectively demonstrated by a project in arid Oklahoma. There a twenty-two-acre field that was as bare of vegetation as a pavement when the project was started, six years later produced sixty-two and one-half bushels of Oregon turf winter oats to the acre, and nothing but limestone and weeds were used to develop the soil. Not one pound of commercial fertilizer was used.

Thus weeds can be controlled by using them; but of course it is much easier to spray them out and use commercial fertilizers. Even then, however, unless a good legume growth is plowed under every three or four years, the humus side of the soil, the moisture-holding side of it, is sure to decrease. Plowing under all residues—weeds, straw, stalks, etc.—and at least one good, well developed legume crop every three or four years, all of them worked into the soil, will keep the organic and living side of the soil in good heart. Then the weeds may be, and should be, sprayed out.

And there are places, many of them, where the spraying out of weeds is the only sensible way to be rid of them. The use of herbicides, as well as the making of them, is in its infancy; but there are already enough data to prove that some of our very worst weeds like Canada thistle (P. 293), Johnson grass (P. 24), Quack grass, Tall foxtail, etc., can be destroyed with spray materials that are so well adjusted to the weed species and to the time of application that a 100% kill may be guaranteed.

Nearly everyone who has trouble with weeds has heard of 2,4–D, and perhaps 2,4,5–T. These numbers and letters have meaning, but the husbandman, the lawn-owner, the gardener, or anyone else who has weed trouble needs to know that these are the broad-leaf and not grass-like weed sprays. And everyone who uses a weed spray should know that directions on the can or bag should be followed. But even if he does know all of these pertinent facts, and follows them to the letter, he may fail to get satisfactory results, just as the cook who follows a recipe to the letter may fail to get the results obtained by the chef. A good technician of any kind has something more than the ability to follow directions. The state of the weather, the time of the day, the makeup of the spray material, these and perhaps many more conditions are sensed by the sprayer who obtains a perfect kill.

On page 37 of the Special Report of the Agricultural Research Service (ARS) of the U. S. Department of Agriculture, the formulae and chemical compounds of herbaceous weed sprays are given. They are under the heading "Herbaceous Perennial Weeds," but the reader must know that the selfsame sprays are used on the annual weeds as on the perennial. Because these formulae and compounds are the very latest, and are herbicides fully tested in use, they are given here exactly as they are given in the ARS report.

HERBACEOUS PERENNIAL WEEDS

Broadleaved Perennials:
2,4–D, MCPA, 2,4,5–T

Perennial Grasses and Sedges:
Cultivated areas—TCA, dalapon, aromatic oils, dinitro-fortified fuel oils; non-cultivated areas—urea compounds such as monuron, diuron, and fenuron, sodium chlorate, borax.

WOODY PLANTS AND WEEDS ALONG FENCEROWS, DITCHBANKS, ROADSIDES, UTILITY LINES, AND ON NON-CULTIVATED AREAS

Woody Plants and Weeds:
2,4-D, 2,4,5-T, 2 (2,4,5-TP), MCPA, ammonium sulfamate

Trees and Brush Less Than 6 Inches in Diameter:
2,4-D, 2,4,5-T, mixtures of the two

Trees 6 Inches in Diameter or Larger:
2,4,5-T

Then on page 38 of this same report the herbicides and chemical compounds known to be effective in soil sterilization are given.

Soil Sterilization:
Sodium chlorate, borax, sodium arsenite, urea herbicides, and mixtures of these herbicides with dalapon, TCA, 2,4,5-T, 2,4-D, herbicidal oils, dinitro compounds, pentachlorophenol, or other fortifying agents.

Before closing this chapter on weed control it might be well to remind the reader that these herbicides are poisonous, not only to weed cells but to other living cells. This fact is usually stated on every can and package containing the chemicals, under *Caution* and *Antidote*. But many people fail to read such trivial things; in fact, many will drop the given amount of chemicals into the given amount of water and without any agitation of the mixture will then spray it on the weeds. The mixture must be thoroughly agitated before spraying. These suggestions are for only those who use hand sprayers. The power sprayers take care of all of this, and the *poison caution* is not so pertinent for those using power sprayers. It is pertinent, however, for those who mix the spray material by stirring it, and especially for those who have inquisitive small children. In the latter case these herbicide packages should be treated as all poison-containing packages should be treated: placed under lock and key.

Among those who read this chapter there are sure to be a few filled with the desire to know just how these herbicides "work." If you happen to know how a plant grows; how it is constructed to take its water requirements from the soil to its sugar-making leaves, and how this sugar, made in the leaves, is transported from the leaves to every living cell in the plant's entire living structure, then you should be able to read and understand Prof. A. S. Craft's excellent article, "Weed Control: Applied Botany" in the July, 1956 issue of the *American Journal of Botany*.

Another valuable source of information on weed control is to be found in the report of the North Central Weed Control Council, the NCWCC, published from Omaha, Nebraska, Dec. 5, 1955. It deals, of course, with the control of weeds indigenous to the North Central States, but most of the weeds of that area are found in somewhat similar areas throughout the United States. The worst weeds seldom "stay put." There are, however, a few that are at their worst in certain areas such as, for instance, Canada thistle (P. 293) and Quack grass in the northern states; Bermuda grass (P. 46) and Johnson grass (P. 24) in the southern areas. But the big majority of weed species fare well wherever there is water and soil enough to support them.

The following paragraph is taken from a most interesting paper, copies of which should be available:

Weeds are among the greatest contributors to production costs on American farms. The losses caused by weeds on farms in the United States have now reached an estimated four billion dollars annually. These losses are estimated to equal the combined losses from insects and diseases, and are second only to farm losses caused by soil erosion.

This paper was written by Warren C. Shaw, Agronomist, Field Crops Research Branch, Agricultural Research Service, United States Department of Agriculture, Plant Industry Station, Beltsville, Maryland, and was delivered by Mr. Shaw at the Second British Weed Control Conference held at Harrogate, England, November 2, 3 and 4, 1954. His report deals with the herbicidal effects of certain chemical compounds, and with the research designed to find effective herbicides; it was intended for the organic chemist who thinks of 2,4-D, for instance, as represented in this way:

But the paragraph quoted above, and the following paragraph taken from near the close of the report, should convince the reader that the grower of plants must needs use every possible means for controlling weeds, and then be satisfied with having been in a war that does give promise of being won some day in some places:

> Of much greater significance and practical importance are the changes in ecological relationships as a result of the use of herbicides. This is not mere speculation as to what may happen. We already have ample proof as measured by what has happened. In the North Central United States, 2,4-D is used for the control of weeds in over 8 million acres of corn annually. The weedy grasses and many serious broadleaved weeds are not controlled by the treatments. As a result the broadleaved annual weeds are decreasing and the grasses are increasing, presenting a different and in many cases a more difficult weed problem than the original. Over one million acres of mesquite, a weedy perennial, and other brush is being treated annually with 2,4,5-T in the Southwestern United States. Many pernicious perennial weedy plants are not killed by 2,4,5-T and these species are on the increase. These are but two examples of shifts in ecological relationships which we must deal with in the future. There will be others. We must design our research programs to be versatile enough to meet the challenges of these ecological changes.

There is no doubt that the day will come when weeds will be under much more control than they have been in the past, but it is quite possible then that Mother Nature will play another card that she has had up her sleeve all the time. The vitamin values of cultivated food crops may fall. They already have fallen in places where the humus content of the soil is permitted to decrease. There is no doubt that trace-elements play a greater part in all forms of life, from mushroom to man, than is yet fully appreciated. Some of our sturdiest weeds evidently tap a trace-element reservoir found in many of our subsoils. The decay of these weeds in our compost pits, and in our top soil when a crop of flourishing weeds is plowed into it, substantiate the claims of the compost maker, and testify to the intelligence of the farmer or gardener who controls his weeds by using them.

BIBLIOGRAPHY

BIBLIOGRAPHY

Aslander, Alfred, *Sulphuric Acid as a Weed Spray*. Jr. Ag. Res. 34: 1065–1091 (June 1, 1927).

Ball, W. E., and French, O. C., *Sulphuric Acid for Control of Weeds*. Uni. Calif. Ex. Sta. Bl. No. 590 (1935).

Beal, W. J., *Michigan Weeds*. Mich. Ag. Coll. Ex. Sta. Bl. No. 267 (1911).

Blatchley, W. S., *Indiana Weed Book*. Nature Publishing Co., Indianapolis, Ind., 1920.

* Britton, Nathaniel Lord, and Brown, Hon. Addison, *An Illustrated Flora of the Northern United States, Canada and the British Possessions*. Three volumes. Charles Scribner's Sons, New York, 1913.

Chestnut, V. K., *Thirty Poisonous Weeds*. U. S. Dept. Ag. Farmers' Bl. No. 28 (1898).

Cox, H. R., *Wild Onion: Methods of Iradication*. U. S. Dept. Ag. Farmers' Bl. No. 610 (Reprint 1922).

—— *The Eradication of Bindweed or Wild Morning Glory*. U. S. Dept. Ag. Farmers' Bl. No. 368 (1915).

Crafts, A. S., *Weed Control: Applied Botany*. American Journal of Botany. July 1956, Vol. 43, No. 7, P. 548–555.

Graham, Robert, and Michael, V. M., *Wild Snakeroot Poisoning*. Univ. Ill. Ag. Ex. Sta. Cir. No. 436 (1935).

Gress, E. M., PhD., *Pennsylvania Weeds*. Penn. Dept. Ag. Bl. No. 20, Harrisburg, Pa., 1925.

* Grieve, Mrs. M., and Leyel, Mrs. C. F., *A Modern Herbal*, Vols. I and II. Harcourt, Brace & Co., New York, 1931.

Griffeth, R. Eglesfeed, M.D., *Medicinal Botany*. Lea & Blanchard, Philadelphia, 1847.

Harvey, F. A., *Three Troublesome Weeds*. Me. State Coll. Ag. Ex. Sta. Bl. No. 32 (1897).

Hinkel, Alice, *Weeds Used in Medicine*. U. S. Dept. Ag. Farmers' Bl. No. 188 (1904).

* Hitchcock, A. S., *Manual of the Grasses of the United States*. U. S. Govt. Printing Office, Washington, 1935.

Latshaw, W. L., and Zahnley, J. W., *Experiments with Sodium Chlorate and Other Chemicals as Herbicides for Field Bindweed*. Jr. Ag. Res. 35:757–767, Oct. 15, 1927.

Long, Harold C., and Percival, John, *Common Weeds of Farm and Garden*. Smith, Elder and Co., London, 1910.

* Dover reprint.

* Millspaugh, Chas. F., M.D., *Medicinal Plants.* Two volumes. John C. Yorston & Co., Philadelphia, 1892.

Muenscher, Walter Conrad, *Weeds.* The Macmillan Company, New York, 1935.

Munson, W. M. *Dandelions, Hawkweeds, Ginseng, Canker Worms,* Me. Ag. Ex. Sta. Bl. No. 95 (1903).

Oswald, W. L., and Boss, Andrew, *Minnesota Weeds* (Series III). Univ. Minn. Ag. Ex. Sta. Bl. No. 176 (1918).

Pammel, L. H., *Weeds of the Farm and Garden.* Orange Judd Co., New York, 1911.

―――― *Unlawful and Other Weeds of Iowa.* Ag. Extension Bl. No. 3, Ames, Iowa, 1915.

Pipal, T. J., *Red Sorrel and Its Control.* Purdue Univ. Ag. Ex. Sta. Bl. No. 197 (1916).

―――― *Wild Garlic and Its Eradication.* Purdue Uni. Ag. Ex. Sta. Bl. No. 176 (1914).

Robinson, Benjamin Lincoln, and Fernald, Merrit Lydon, *Gray's Manual of Botany.* 7th Edition, Illustrated, American Book Company, New York, Cincinnati, Chicago, 1908.

Scott, O. M. and Sons Co., *Campus and Athletic Field.* Marysville, Ohio, 1932.

―――― *Lawn Care.* Pamphlets beginning 1928――――.

Shaw, Warren C., Agronomist, U. S. Dept. Ag., Beltsville, Md. a paper "*Recent Advances in Weed Control in the United States*" delivered at the Second Weed Control Conference, Harrogate, England, Nov. 1954.

Stockberger, W. W., *Drug Plants Under Cultivation.* U. S. Dept. Ag. Farmers' Bl. No. 663 (1915).

Suggested Guide for Chemical Control of Weeds, April, 1956. Agricultural Research Service. U. S. Dept. of Ag.

Talbot, M. W., *Johnson Grass as a Weed.* U. S. Dept. Ag. Farmers' Bl. No. 1537 (1928).

Tehon, L. R., *Rout the Weeds, Why, When and How.* Ill. Nat. Hist. Sur. Cir. No. 28, Urbana, Ill., Aug., 1937.

The Oxford English Dictionary, Edited by James A. H. Murray, Henry Bradley, W. A. Craigie, and C. T. Onions, At the Clarendon Press, Oxford, 1932.

U. S. Dispensatories (Several editions).

U. S. Pharmacopoeias (Several editions).

Webster's New International Dictionary, 2d Edition, Unabridged.

―――――――――――――

* Dover reprint.

INDEX

INDEX

A CATALOG OF SELECTED
DOVER BOOKS
IN ALL FIELDS OF INTEREST

A CATALOG OF SELECTED DOVER
BOOKS IN ALL FIELDS OF INTEREST

DRAWINGS OF REMBRANDT, edited by Seymour Slive. Updated Lippmann, Hofstede de Groot edition, with definitive scholarly apparatus. All portraits, biblical sketches, landscapes, nudes. Oriental figures, classical studies, together with selection of work by followers. 550 illustrations. Total of 630pp. 9⅛ × 12¼.
21485-0, 21486-9 Pa., Two-vol. set $29.90

GHOST AND HORROR STORIES OF AMBROSE BIERCE, Ambrose Bierce. 24 tales vividly imagined, strangely prophetic, and decades ahead of their time in technical skill: "The Damned Thing," "An Inhabitant of Carcosa," "The Eyes of the Panther," "Moxon's Master," and 20 more. 199pp. 5⅜ × 8½. 20767-6 Pa. $3.95

ETHICAL WRITINGS OF MAIMONIDES, Maimonides. Most significant ethical works of great medieval sage, newly translated for utmost precision, readability. Laws Concerning Character Traits, Eight Chapters, more. 192pp. 5⅜ × 8½.
24522-5 Pa. $4.50

THE EXPLORATION OF THE COLORADO RIVER AND ITS CANYONS, J. W. Powell. Full text of Powell's 1,000-mile expedition down the fabled Colorado in 1869. Superb account of terrain, geology, vegetation, Indians, famine, mutiny, treacherous rapids, mighty canyons, during exploration of last unknown part of continental U.S. 400pp. 5⅜ × 8½. 20094-9 Pa. $7.95

HISTORY OF PHILOSOPHY, Julián Marías. Clearest one-volume history on the market. Every major philosopher and dozens of others, to Existentialism and later. 505pp. 5⅜ × 8½. 21739-6 Pa. $9.95

ALL ABOUT LIGHTNING, Martin A. Uman. Highly readable non-technical survey of nature and causes of lightning, thunderstorms, ball lightning, St. Elmo's Fire, much more. Illustrated. 192pp. 5⅜ × 8½. 25237-X Pa. $5.95

SAILING ALONE AROUND THE WORLD, Captain Joshua Slocum. First man to sail around the world, alone, in small boat. One of great feats of seamanship told in delightful manner. 67 illustrations. 294pp. 5⅜ × 8½. 20326-3 Pa. $4.95

LETTERS AND NOTES ON THE MANNERS, CUSTOMS AND CONDI-TIONS OF THE NORTH AMERICAN INDIANS, George Catlin. Classic account of life among Plains Indians: ceremonies, hunt, warfare, etc. 312 plates. 572pp. of text. 6⅛ × 9¼. 22118-0, 22119-9, Pa. Two-vol. set $17.90

ALASKA: The Harriman Expedition, 1899, John Burroughs, John Muir, et al. Informative, engrossing accounts of two-month, 9,000-mile expedition. Native peoples, wildlife, forests, geography, salmon industry, glaciers, more. Profusely illustrated. 240 black-and-white line drawings. 124 black-and-white photographs. 3 maps. Index. 576pp. 5⅜ × 8½. 25109-8 Pa. $11.95

THE BOOK OF BEASTS: Being a Translation from a Latin Bestiary of the Twelfth Century, T. H. White. Wonderful catalog real and fanciful beasts: manticore, griffin, phoenix, amphivius, jaculus, many more. White's witty erudite commentary on scientific, historical aspects. Fascinating glimpse of medieval mind. Illustrated. 296pp. 5⅜ × 8¼. (Available in U.S. only)   24609-4 Pa. $6.95

FRANK LLOYD WRIGHT: ARCHITECTURE AND NATURE With 160 Illustrations, Donald Hoffmann. Profusely illustrated study of influence of nature—especially prairie—on Wright's designs for Fallingwater, Robie House, Guggenheim Museum, other masterpieces. 96pp. 9¼ × 10¾.   25098-9 Pa. $7.95

FRANK LLOYD WRIGHT'S FALLINGWATER, Donald Hoffmann. Wright's famous waterfall house: planning and construction of organic idea. History of site, owners, Wright's personal involvement. Photographs of various stages of building. Preface by Edgar Kaufmann, Jr. 100 illustrations. 112pp. 9¼ × 10.
23671-4 Pa. $8.95

YEARS WITH FRANK LLOYD WRIGHT: Apprentice to Genius, Edgar Tafel. Insightful memoir by a former apprentice presents a revealing portrait of Wright the man, the inspired teacher, the greatest American architect. 372 black-and-white illustrations. Preface. Index. vi + 228pp. 8¼ × 11.   24801-1 Pa. $10.95

THE STORY OF KING ARTHUR AND HIS KNIGHTS, Howard Pyle. Enchanting version of King Arthur fable has delighted generations with imaginative narratives of exciting adventures and unforgettable illustrations by the author. 41 illustrations. xviii + 313pp. 6⅛ × 9¼.   21445-1 Pa. $6.95

THE GODS OF THE EGYPTIANS, E. A. Wallis Budge. Thorough coverage of numerous gods of ancient Egypt by foremost Egyptologist. Information on evolution of cults, rites and gods; the cult of Osiris; the Book of the Dead and its rites; the sacred animals and birds; Heaven and Hell; and more. 956pp. 6⅛ × 9¼.
22055-9, 22056-7 Pa., Two-vol. set $21.90

A THEOLOGICO-POLITICAL TREATISE, Benedict Spinoza. Also contains unfinished *Political Treatise*. Great classic on religious liberty, theory of government on common consent. R. Elwes translation. Total of 421pp. 5⅜ × 8½.
20249-6 Pa. $6.95

INCIDENTS OF TRAVEL IN CENTRAL AMERICA, CHIAPAS, AND YUCATAN, John L. Stephens. Almost single-handed discovery of Maya culture; exploration of ruined cities, monuments, temples; customs of Indians. 115 drawings. 892pp. 5⅜ × 8½.   22404-X, 22405-8 Pa., Two-vol. set $15.90

LOS CAPRICHOS, Francisco Goya. 80 plates of wild, grotesque monsters and caricatures. Prado manuscript included. 183pp. 6⅜ × 9⅜.   22384-1 Pa. $5.95

AUTOBIOGRAPHY: The Story of My Experiments with Truth, Mohandas K. Gandhi. Not hagiography, but Gandhi in his own words. Boyhood, legal studies, purification, the growth of the Satyagraha (nonviolent protest) movement. Critical, inspiring work of the man who freed India. 480pp. 5⅜ × 8½. (Available in U.S. only)
24593-4 Pa. $6.95

ILLUSTRATED DICTIONARY OF HISTORIC ARCHITECTURE, edited by Cyril M. Harris. Extraordinary compendium of clear, concise definitions for over 5,000 important architectural terms complemented by over 2,000 line drawings. Covers full spectrum of architecture from ancient ruins to 20th-century Modernism. Preface. 592pp. 7½ × 9⅝. 24444-X Pa. $15.95

THE NIGHT BEFORE CHRISTMAS, Clement Moore. Full text, and woodcuts from original 1848 book. Also critical, historical material. 19 illustrations. 40pp. 4⅝ × 6. 22797-9 Pa. $2.50

THE LESSON OF JAPANESE ARCHITECTURE: 165 Photographs, Jiro Harada. Memorable gallery of 165 photographs taken in the 1930's of exquisite Japanese homes of the well-to-do and historic buildings. 13 line diagrams. 192pp. 8⅝ × 11¼. 24778-3 Pa. $10.95

THE AUTOBIOGRAPHY OF CHARLES DARWIN AND SELECTED LETTERS, edited by Francis Darwin. The fascinating life of eccentric genius composed of an intimate memoir by Darwin (intended for his children); commentary by his son, Francis; hundreds of fragments from notebooks, journals, papers; and letters to and from Lyell, Hooker, Huxley, Wallace and Henslow. xi + 365pp. 5⅜ × 8. 20479-0 Pa. $6.95

WONDERS OF THE SKY: Observing Rainbows, Comets, Eclipses, the Stars and Other Phenomena, Fred Schaaf. Charming, easy-to-read poetic guide to all manner of celestial events visible to the naked eye. Mock suns, glories, Belt of Venus, more. Illustrated. 299pp. 5¼ × 8¼. 24402-4 Pa. $7.95

BURNHAM'S CELESTIAL HANDBOOK, Robert Burnham, Jr. Thorough guide to the stars beyond our solar system. Exhaustive treatment. Alphabetical by constellation: Andromeda to Cetus in Vol. 1; Chamaeleon to Orion in Vol. 2; and Pavo to Vulpecula in Vol. 3. Hundreds of illustrations. Index in Vol. 3. 2,000pp. 6⅛ × 9¼. 23567-X, 23568-8, 23673-0 Pa., Three-vol. set $38.85

STAR NAMES: Their Lore and Meaning, Richard Hinckley Allen. Fascinating history of names various cultures have given to constellations and literary and folkloristic uses that have been made of stars. Indexes to subjects. Arabic and Greek names. Biblical references. Bibliography. 563pp. 5⅜ × 8½. 21079-0 Pa. $8.95

THIRTY YEARS THAT SHOOK PHYSICS: The Story of Quantum Theory, George Gamow. Lucid, accessible introduction to influential theory of energy and matter. Careful explanations of Dirac's anti-particles, Bohr's model of the atom, much more. 12 plates. Numerous drawings. 240pp. 5⅜ × 8½. 24895-X Pa. $5.95

CHINESE DOMESTIC FURNITURE IN PHOTOGRAPHS AND MEASURED DRAWINGS, Gustav Ecke. A rare volume, now affordably priced for antique collectors, furniture buffs and art historians. Detailed review of styles ranging from early Shang to late Ming. Unabridged republication. 161 black-and-white drawings, photos. Total of 224pp. 8⅜ × 11¼. (Available in U.S. only) 25171-3 Pa. $13.95

VINCENT VAN GOGH: A Biography, Julius Meier-Graefe. Dynamic, penetrating study of artist's life, relationship with brother, Theo, painting techniques, travels, more. Readable, engrossing. 160pp. 5⅜ × 8½. (Available in U.S. only) 25253-1 Pa. $4.95

HOW TO WRITE, Gertrude Stein. Gertrude Stein claimed anyone could understand her unconventional writing—here are clues to help. Fascinating improvisations, language experiments, explanations illuminate Stein's craft and the art of writing. Total of 414pp. 4⅜ × 6⅜. 23144-5 Pa. $6.95

ADVENTURES AT SEA IN THE GREAT AGE OF SAIL: Five Firsthand Narratives, edited by Elliot Snow. Rare true accounts of exploration, whaling, shipwreck, fierce natives, trade, shipboard life, more. 33 illustrations. Introduction. 353pp. 5⅜ × 8½. 25177-2 Pa. $8.95

THE HERBAL OR GENERAL HISTORY OF PLANTS, John Gerard. Classic descriptions of about 2,850 plants—with over 2,700 illustrations—includes Latin and English names, physical descriptions, varieties, time and place of growth, more. 2,706 illustrations. xlv + 1,678pp. 8½ × 12¼. 23147-X Cloth. $75.00

DOROTHY AND THE WIZARD IN OZ, L. Frank Baum. Dorothy and the Wizard visit the center of the Earth, where people are vegetables, glass houses grow and Oz characters reappear. Classic sequel to *Wizard of Oz*. 256pp. 5⅜ × 8. 24714-7 Pa. $4.95

SONGS OF EXPERIENCE: Facsimile Reproduction with 26 Plates in Full Color, William Blake. This facsimile of Blake's original "Illuminated Book" reproduces 26 full-color plates from a rare 1826 edition. Includes "The Tyger," "London," "Holy Thursday," and other immortal poems. 26 color plates. Printed text of poems. 48pp. 5¼ × 7. 24636-1 Pa. $3.50

SONGS OF INNOCENCE, William Blake. The first and most popular of Blake's famous "Illuminated Books," in a facsimile edition reproducing all 31 brightly colored plates. Additional printed text of each poem. 64pp. 5¼ × 7. 22764-2 Pa. $3.50

PRECIOUS STONES, Max Bauer. Classic, thorough study of diamonds, rubies, emeralds, garnets, etc.: physical character, occurrence, properties, use, similar topics. 20 plates, 8 in color. 94 figures. 659pp. 6⅛ × 9¼. 21910-0, 21911-9 Pa., Two-vol. set $15.90

ENCYCLOPEDIA OF VICTORIAN NEEDLEWORK, S. F. A. Caulfeild and Blanche Saward. Full, precise descriptions of stitches, techniques for dozens of needlecrafts—most exhaustive reference of its kind. Over 800 figures. Total of 679pp. 8⅛ × 11. Two volumes. Vol. 1 22800-2 Pa. $11.95
Vol. 2 22801-0 Pa. $11.95

THE MARVELOUS LAND OF OZ, L. Frank Baum. Second Oz book, the Scarecrow and Tin Woodman are back with hero named Tip, Oz magic. 136 illustrations. 287pp. 5⅜ × 8½. 20692-0 Pa. $5.95

WILD FOWL DECOYS, Joel Barber. Basic book on the subject, by foremost authority and collector. Reveals history of decoy making and rigging, place in American culture, different kinds of decoys, how to make them, and how to use them. 140 plates. 156pp. 7⅞ × 10¾. 20011-6 Pa. $8.95

HISTORY OF LACE, Mrs. Bury Palliser. Definitive, profusely illustrated chronicle of lace from earliest times to late 19th century. Laces of Italy, Greece, England, France, Belgium, etc. Landmark of needlework scholarship. 266 illustrations. 672pp. 6⅛ × 9¼. 24742-2 Pa. $14.95

ILLUSTRATED GUIDE TO SHAKER FURNITURE, Robert Meader. All furniture and appurtenances, with much on unknown local styles. 235 photos. 146pp. 9 × 12.   22819-3 Pa. $8.95

WHALE SHIPS AND WHALING: A Pictorial Survey, George Francis Dow. Over 200 vintage engravings, drawings, photographs of barks, brigs, cutters, other vessels. Also harpoons, lances, whaling guns, many other artifacts. Comprehensive text by foremost authority. 207 black-and-white illustrations. 288pp. 6 × 9.
                                                                    24808-9 Pa. $8.95

THE BERTRAMS, Anthony Trollope. Powerful portrayal of blind self-will and thwarted ambition includes one of Trollope's most heartrending love stories. 497pp. 5⅜ × 8½.   25119-5 Pa. $9.95

ADVENTURES WITH A HAND LENS, Richard Headstrom. Clearly written guide to observing and studying flowers and grasses, fish scales, moth and insect wings, egg cases, buds, feathers, seeds, leaf scars, moss, molds, ferns, common crystals, etc.—all with an ordinary, inexpensive magnifying glass. 209 exact line drawings aid in your discoveries. 220pp. 5⅜ × 8½.   23330-8 Pa. $4.95

RODIN ON ART AND ARTISTS, Auguste Rodin. Great sculptor's candid, wide-ranging comments on meaning of art; great artists; relation of sculpture to poetry, painting, music; philosophy of life, more. 76 superb black-and-white illustrations of Rodin's sculpture, drawings and prints. 119pp. 8⅜ × 11¼.   24487-3 Pa. $7.95

FIFTY CLASSIC FRENCH FILMS, 1912–1982: A Pictorial Record, Anthony Slide. Memorable stills from Grand Illusion, Beauty and the Beast, Hiroshima, Mon Amour, many more. Credits, plot synopses, reviews, etc. 160pp. 8¼ × 11.
                                                                    25256-6 Pa. $11.95

THE PRINCIPLES OF PSYCHOLOGY, William James. Famous long course complete, unabridged. Stream of thought, time perception, memory, experimental methods; great work decades ahead of its time. 94 figures. 1,391pp. 5⅜ × 8½.
                                        20381-6, 20382-4 Pa., Two-vol. set $23.90

BODIES IN A BOOKSHOP, R. T. Campbell. Challenging mystery of blackmail and murder with ingenious plot and superbly drawn characters. In the best tradition of British suspense fiction. 192pp. 5⅜ × 8½.   24720-1 Pa. $3.95

CALLAS: PORTRAIT OF A PRIMA DONNA, George Jellinek. Renowned commentator on the musical scene chronicles incredible career and life of the most controversial, fascinating, influential operatic personality of our time. 64 black-and-white photographs. 416pp. 5⅜ × 8¼.   25047-4 Pa. $8.95

GEOMETRY, RELATIVITY AND THE FOURTH DIMENSION, Rudolph Rucker. Exposition of fourth dimension, concepts of relativity as Flatland characters continue adventures. Popular, easily followed yet accurate, profound. 141 illustrations. 133pp. 5⅜ × 8½.   23400-2 Pa. $3.95

HOUSEHOLD STORIES BY THE BROTHERS GRIMM, with pictures by Walter Crane. 53 classic stories—Rumpelstiltskin, Rapunzel, Hansel and Gretel, the Fisherman and his Wife, Snow White, Tom Thumb, Sleeping Beauty, Cinderella, and so much more—lavishly illustrated with original 19th century drawings. 114 illustrations. x + 269pp. 5⅜ × 8½.   21080-4 Pa. $4.95

CATALOG OF DOVER BOOKS

SUNDIALS, Albert Waugh. Far and away the best, most thorough coverage of ideas, mathematics concerned, types, construction, adjusting anywhere. Over 100 illustrations. 230pp. 5⅜ × 8½. 22947-5 Pa. $4.95

PICTURE HISTORY OF THE NORMANDIE: With 190 Illustrations, Frank O. Braynard. Full story of legendary French ocean liner: Art Deco interiors, design innovations, furnishings, celebrities, maiden voyage, tragic fire, much more. Extensive text. 144pp. 8⅞ × 11¾. 25257-4 Pa. $10.95

THE FIRST AMERICAN COOKBOOK: A Facsimile of "American Cookery," 1796, Amelia Simmons. Facsimile of the first American-written cookbook published in the United States contains authentic recipes for colonial favorites—pumpkin pudding, winter squash pudding, spruce beer, Indian slapjacks, and more. Introductory Essay and Glossary of colonial cooking terms. 80pp. 5⅜ × 8½. 24710-4 Pa. $3.50

101 PUZZLES IN THOUGHT AND LOGIC, C. R. Wylie, Jr. Solve murders and robberies, find out which fishermen are liars, how a blind man could possibly identify a color—purely by your own reasoning! 107pp. 5⅜ × 8½. 20367-0 Pa. $2.50

THE BOOK OF WORLD-FAMOUS MUSIC—CLASSICAL, POPULAR AND FOLK, James J. Fuld. Revised and enlarged republication of landmark work in musico-bibliography. Full information about nearly 1,000 songs and compositions including first lines of music and lyrics. New supplement. Index. 800pp. 5⅜ × 8¼. 24857-7 Pa. $15.95

ANTHROPOLOGY AND MODERN LIFE, Franz Boas. Great anthropologist's classic treatise on race and culture. Introduction by Ruth Bunzel. Only inexpensive paperback edition. 255pp. 5⅜ × 8½. 25245-0 Pa. $6.95

THE TALE OF PETER RABBIT, Beatrix Potter. The inimitable Peter's terrifying adventure in Mr. McGregor's garden, with all 27 wonderful, full-color Potter illustrations. 55pp. 4¼ × 5½. (Available in U.S. only) 22827-4 Pa. $1.75

THREE PROPHETIC SCIENCE FICTION NOVELS, H. G. Wells. *When the Sleeper Wakes, A Story of the Days to Come* and *The Time Machine* (full version). 335pp. 5⅜ × 8½. (Available in U.S. only) 20605-X Pa. $6.95

APICIUS COOKERY AND DINING IN IMPERIAL ROME, edited and translated by Joseph Dommers Vehling. Oldest known cookbook in existence offers readers a clear picture of what foods Romans ate, how they prepared them, etc. 49 illustrations. 301pp. 6⅛ × 9¼. 23563-7 Pa. $7.95

SHAKESPEARE LEXICON AND QUOTATION DICTIONARY, Alexander Schmidt. Full definitions, locations, shades of meaning of every word in plays and poems. More than 50,000 exact quotations. 1,485pp. 6½ × 9¼. 22726-X, 22727-8 Pa., Two-vol. set $29.90

THE WORLD'S GREAT SPEECHES, edited by Lewis Copeland and Lawrence W. Lamm. Vast collection of 278 speeches from Greeks to 1970. Powerful and effective models; unique look at history. 842pp. 5⅜ × 8½. 20468-5 Pa. $11.95

THE BLUE FAIRY BOOK, Andrew Lang. The first, most famous collection, with many familiar tales: Little Red Riding Hood, Aladdin and the Wonderful Lamp, Puss in Boots, Sleeping Beauty, Hansel and Gretel, Rumpelstiltskin; 37 in all. 138 illustrations. 390pp. 5⅜ × 8½. 21437-0 Pa. $6.95

THE STORY OF THE CHAMPIONS OF THE ROUND TABLE, Howard Pyle. Sir Launcelot, Sir Tristram and Sir Percival in spirited adventures of love and triumph retold in Pyle's inimitable style. 50 drawings, 31 full-page. xviii + 329pp. 6½ × 9¼. 21883-X Pa. $7.95

AUDUBON AND HIS JOURNALS, Maria Audubon. Unmatched two-volume portrait of the great artist, naturalist and author contains his journals, an excellent biography by his granddaughter, expert annotations by the noted ornithologist, Dr. Elliott Coues, and 37 superb illustrations. Total of 1,200pp. 5⅜ × 8.

Vol. I 25143-8 Pa. $8.95
Vol. II 25144-6 Pa. $8.95

GREAT DINOSAUR HUNTERS AND THEIR DISCOVERIES, Edwin H. Colbert. Fascinating, lavishly illustrated chronicle of dinosaur research, 1820's to 1960. Achievements of Cope, Marsh, Brown, Buckland, Mantell, Huxley, many others. 384pp. 5¼ × 8¼. 24701-5 Pa. $7.95

THE TASTEMAKERS, Russell Lynes. Informal, illustrated social history of American taste 1850's-1950's. First popularized categories Highbrow, Lowbrow, Middlebrow. 129 illustrations. New (1979) afterword. 384pp. 6 × 9. 23993-4 Pa. $8.95

DOUBLE CROSS PURPOSES, Ronald A. Knox. A treasure hunt in the Scottish Highlands, an old map, unidentified corpse, surprise discoveries keep reader guessing in this cleverly intricate tale of financial skullduggery. 2 black-and-white maps. 320pp. 5⅜ × 8½. (Available in U.S. only) 25032-6 Pa. $6.95

AUTHENTIC VICTORIAN DECORATION AND ORNAMENTATION IN FULL COLOR: 46 Plates from "Studies in Design," Christopher Dresser. Superb full-color lithographs reproduced from rare original portfolio of a major Victorian designer. 48pp. 9¼ × 12¼. 25083-0 Pa. $7.95

PRIMITIVE ART, Franz Boas. Remains the best text ever prepared on subject, thoroughly discussing Indian, African, Asian, Australian, and, especially, Northern American primitive art. Over 950 illustrations show ceramics, masks, totem poles, weapons, textiles, paintings, much more. 376pp. 5⅜ × 8. 20025-6 Pa. $6.95

SIDELIGHTS ON RELATIVITY, Albert Einstein. Unabridged republication of two lectures delivered by the great physicist in 1920-21. *Ether and Relativity* and *Geometry and Experience*. Elegant ideas in non-mathematical form, accessible to intelligent layman. vi + 56pp. 5⅜ × 8½. 24511-X Pa. $2.95

THE WIT AND HUMOR OF OSCAR WILDE, edited by Alvin Redman. More than 1,000 ripostes, paradoxes, wisecracks: Work is the curse of the drinking classes, I can resist everything except temptation, etc. 258pp. 5⅜ × 8½. 20602-5 Pa. $4.95

ADVENTURES WITH A MICROSCOPE, Richard Headstrom. 59 adventures with clothing fibers, protozoa, ferns and lichens, roots and leaves, much more. 142 illustrations. 232pp. 5⅜ × 8½. 23471-1 Pa. $3.95

PLANTS OF THE BIBLE, Harold N. Moldenke and Alma L. Moldenke. Standard reference to all 230 plants mentioned in Scriptures. Latin name, biblical reference, uses, modern identity, much more. Unsurpassed encyclopedic resource for scholars, botanists, nature lovers, students of Bible. Bibliography. Indexes. 123 black-and-white illustrations. 384pp. 6 × 9. 25069-5 Pa. $8.95

FAMOUS AMERICAN WOMEN: A Biographical Dictionary from Colonial Times to the Present, Robert McHenry, ed. From Pocahontas to Rosa Parks, 1,035 distinguished American women documented in separate biographical entries. Accurate, up-to-date data, numerous categories, spans 400 years. Indices. 493pp. 6½ × 9¼. 24523-3 Pa. $10.95

THE FABULOUS INTERIORS OF THE GREAT OCEAN LINERS IN HISTORIC PHOTOGRAPHS, William H. Miller, Jr. Some 200 superb photographs capture exquisite interiors of world's great "floating palaces"—1890's to 1980's: *Titanic, Ile de France, Queen Elizabeth, United States, Europa,* more. Approx. 200 black-and-white photographs. Captions. Text. Introduction. 160pp. 8⅞ × 11¾. 24756-2 Pa. $9.95

THE GREAT LUXURY LINERS, 1927–1954: A Photographic Record, William H. Miller, Jr. Nostalgic tribute to heyday of ocean liners. 186 photos of Ile de France, Normandie, Leviathan, Queen Elizabeth, United States, many others. Interior and exterior views. Introduction. Captions. 160pp. 9 × 12. 24056-8 Pa. $10.95

A NATURAL HISTORY OF THE DUCKS, John Charles Phillips. Great landmark of ornithology offers complete detailed coverage of nearly 200 species and subspecies of ducks: gadwall, sheldrake, merganser, pintail, many more. 74 full-color plates, 102 black-and-white. Bibliography. Total of 1,920pp. 8⅝ × 11¼. 25141-1, 25142-X Cloth. Two-vol. set $100.00

THE SEAWEED HANDBOOK: An Illustrated Guide to Seaweeds from North Carolina to Canada, Thomas F. Lee. Concise reference covers 78 species. Scientific and common names, habitat, distribution, more. Finding keys for easy identification. 224pp. 5⅜ × 8½. 25215-9 Pa. $6.95

THE TEN BOOKS OF ARCHITECTURE: The 1755 Leoni Edition, Leon Battista Alberti. Rare classic helped introduce the glories of ancient architecture to the Renaissance. 68 black-and-white plates. 336pp. 8⅜ × 11¼. 25239-6 Pa. $14.95

MISS MACKENZIE, Anthony Trollope. Minor masterpieces by Victorian master unmasks many truths about life in 19th-century England. First inexpensive edition in years. 392pp. 5⅜ × 8½. 25201-9 Pa. $8.95

THE RIME OF THE ANCIENT MARINER, Gustave Doré, Samuel Taylor Coleridge. Dramatic engravings considered by many to be his greatest work. The terrifying space of the open sea, the storms and whirlpools of an unknown ocean, the ice of Antarctica, more—all rendered in a powerful, chilling manner. Full text. 38 plates. 77pp. 9¼ × 12. 22305-1 Pa. $4.95

THE EXPEDITIONS OF ZEBULON MONTGOMERY PIKE, Zebulon Montgomery Pike. Fascinating first-hand accounts (1805–6) of exploration of Mississippi River, Indian wars, capture by Spanish dragoons, much more. 1,088pp. 5⅜ × 8½. 25254-X, 25255-8 Pa. Two-vol. set $25.90

A CONCISE HISTORY OF PHOTOGRAPHY: Third Revised Edition, Helmut Gernsheim. Best one-volume history—camera obscura, photochemistry, daguerreotypes, evolution of cameras, film, more. Also artistic aspects—landscape, portraits, fine art, etc. 281 black-and-white photographs. 26 in color. 176pp. 8⅜ × 11¼. 25128-4 Pa. $13.95

THE DORÉ BIBLE ILLUSTRATIONS, Gustave Doré. 241 detailed plates from the Bible: the Creation scenes, Adam and Eve, Flood, Babylon, battle sequences, life of Jesus, etc. Each plate is accompanied by the verses from the King James version of the Bible. 241pp. 9 × 12. 23004-X Pa. $9.95

HUGGER-MUGGER IN THE LOUVRE, Elliot Paul. Second Homer Evans mystery-comedy. Theft at the Louvre involves sleuth in hilarious, madcap caper. "A knockout."—Books. 336pp. 5⅜ × 8½. 25185-3 Pa. $5.95

FLATLAND, E. A. Abbott. Intriguing and enormously popular science-fiction classic explores the complexities of trying to survive as a two-dimensional being in a three-dimensional world. Amusingly illustrated by the author. 16 illustrations. 103pp. 5⅜ × 8½. 20001-9 Pa. $2.50

THE HISTORY OF THE LEWIS AND CLARK EXPEDITION, Meriwether Lewis and William Clark, edited by Elliott Coues. Classic edition of Lewis and Clark's day-by-day journals that later became the basis for U.S. claims to Oregon and the West. Accurate and invaluable geographical, botanical, biological, meteorological and anthropological material. Total of 1,508pp. 5⅜ × 8½.
21268-8, 21269-6, 21270-X Pa. Three-vol. set $26.85

LANGUAGE, TRUTH AND LOGIC, Alfred J. Ayer. Famous, clear introduction to Vienna, Cambridge schools of Logical Positivism. Role of philosophy, elimination of metaphysics, nature of analysis, etc. 160pp. 5⅜ × 8½. (Available in U.S. and Canada only) 20010-8 Pa. $3.95

MATHEMATICS FOR THE NONMATHEMATICIAN, Morris Kline. Detailed, college-level treatment of mathematics in cultural and historical context, with numerous exercises. For liberal arts students. Preface. Recommended Reading Lists. Tables. Index. Numerous black-and-white figures. xvi + 641pp. 5⅜ × 8½.
24823-2 Pa. $11.95

HANDBOOK OF PICTORIAL SYMBOLS, Rudolph Modley. 3,250 signs and symbols, many systems in full; official or heavy commercial use. Arranged by subject. Most in Pictorial Archive series. 143pp. 8⅛ × 11. 23357-X Pa. $6.95

INCIDENTS OF TRAVEL IN YUCATAN, John L. Stephens. Classic (1843) exploration of jungles of Yucatan, looking for evidences of Maya civilization. Travel adventures, Mexican and Indian culture, etc. Total of 669pp. 5⅜ × 8½.
20926-1, 20927-X Pa., Two-vol. set $11.90

CATALOG OF DOVER BOOKS

DEGAS: An Intimate Portrait, Ambroise Vollard. Charming, anecdotal memoir by famous art dealer of one of the greatest 19th-century French painters. 14 black-and-white illustrations. Introduction by Harold L. Van Doren. 96pp. 5⅜ × 8½.
25131-4 Pa. $4.95

PERSONAL NARRATIVE OF A PILGRIMAGE TO ALMANDINAH AND MECCAH, Richard Burton. Great travel classic by remarkably colorful personality. Burton, disguised as a Moroccan, visited sacred shrines of Islam, narrowly escaping death. 47 illustrations. 959pp. 5⅜ × 8½. 21217-3, 21218-1 Pa., Two-vol. set $19.90

PHRASE AND WORD ORIGINS, A. H. Holt. Entertaining, reliable, modern study of more than 1,200 colorful words, phrases, origins and histories. Much unexpected information. 254pp. 5⅜ × 8½. 20758-7 Pa. $5.95

THE RED THUMB MARK, R. Austin Freeman. In this first Dr. Thorndyke case, the great scientific detective draws fascinating conclusions from the nature of a single fingerprint. Exciting story, authentic science. 320pp. 5⅜ × 8½. (Available in U.S. only) 25210-8 Pa. $6.95

AN EGYPTIAN HIEROGLYPHIC DICTIONARY, E. A. Wallis Budge. Monumental work containing about 25,000 words or terms that occur in texts ranging from 3000 B.C. to 600 A.D. Each entry consists of a transliteration of the word, the word in hieroglyphs, and the meaning in English. 1,314pp. 6⅜ × 10. 23615-3, 23616-1 Pa., Two-vol. set $31.90

THE COMPLEAT STRATEGYST: Being a Primer on the Theory of Games of Strategy, J. D. Williams. Highly entertaining classic describes, with many illustrated examples, how to select best strategies in conflict situations. Prefaces. Appendices. xvi + 268pp. 5⅜ × 8½. 25101-2 Pa. $5.95

THE ROAD TO OZ, L. Frank Baum. Dorothy meets the Shaggy Man, little Button-Bright and the Rainbow's beautiful daughter in this delightful trip to the magical Land of Oz. 272pp. 5⅜ × 8. 25208-6 Pa. $5.95

POINT AND LINE TO PLANE, Wassily Kandinsky. Seminal exposition of role of point, line, other elements in non-objective painting. Essential to understanding 20th-century art. 127 illustrations. 192pp. 6½ × 9¼. 23808-3 Pa. $4.95

LADY ANNA, Anthony Trollope. Moving chronicle of Countess Lovel's bitter struggle to win for herself and daughter Anna their rightful rank and fortune—perhaps at cost of sanity itself. 384pp. 5⅜ × 8½. 24669-8 Pa. $8.95

EGYPTIAN MAGIC, E. A. Wallis Budge. Sums up all that is known about magic in Ancient Egypt: the role of magic in controlling the gods, powerful amulets that warded off evil spirits, scarabs of immortality, use of wax images, formulas and spells, the secret name, much more. 253pp. 5⅜ × 8½. 22681-6 Pa. $4.50

THE DANCE OF SIVA, Ananda Coomaraswamy. Preeminent authority unfolds the vast metaphysic of India: the revelation of her art, conception of the universe, social organization, etc. 27 reproductions of art masterpieces. 192pp. 5⅜ × 8½. 24817-8 Pa. $5.95

CATALOG OF DOVER BOOKS

CHRISTMAS CUSTOMS AND TRADITIONS, Clement A. Miles. Origin, evolution, significance of religious, secular practices. Caroling, gifts, yule logs, much more. Full, scholarly yet fascinating; non-sectarian. 400pp. 5⅜ × 8½.
23354-5 Pa. $6.95

THE HUMAN FIGURE IN MOTION, Eadweard Muybridge. More than 4,500 stopped-action photos, in action series, showing undraped men, women, children jumping, lying down, throwing, sitting, wrestling, carrying, etc. 390pp. 7⅞ × 10⅞.
20204-6 Cloth. $21.95

THE MAN WHO WAS THURSDAY, Gilbert Keith Chesterton. Witty, fast-paced novel about a club of anarchists in turn-of-the-century London. Brilliant social, religious, philosophical speculations. 128pp. 5⅜ × 8½.
25121-7 Pa. $3.95

A CEZANNE SKETCHBOOK: Figures, Portraits, Landscapes and Still Lifes, Paul Cezanne. Great artist experiments with tonal effects, light, mass, other qualities in over 100 drawings. A revealing view of developing master painter, precursor of Cubism. 102 black-and-white illustrations. 144pp. 8¾ × 6⅜.
24790-2 Pa. $5.95

AN ENCYCLOPEDIA OF BATTLES: Accounts of Over 1,560 Battles from 1479 B.C. to the Present, David Eggenberger. Presents essential details of every major battle in recorded history, from the first battle of Megiddo in 1479 B.C. to Grenada in 1984. List of Battle Maps. New Appendix covering the years 1967–1984. Index. 99 illustrations. 544pp. 6½ × 9¼.
24913-1 Pa. $14.95

AN ETYMOLOGICAL DICTIONARY OF MODERN ENGLISH, Ernest Weekley. Richest, fullest work, by foremost British lexicographer. Detailed word histories. Inexhaustible. Total of 856pp. 6½ × 9¼.
21873-2, 21874-0 Pa., Two-vol. set $17.00

WEBSTER'S AMERICAN MILITARY BIOGRAPHIES, edited by Robert McHenry. Over 1,000 figures who shaped 3 centuries of American military history. Detailed biographies of Nathan Hale, Douglas MacArthur, Mary Hallaren, others. Chronologies of engagements, more. Introduction. Addenda. 1,033 entries in alphabetical order. xi + 548pp. 6½ × 9¼. (Available in U.S. only)
24758-9 Pa. $13.95

LIFE IN ANCIENT EGYPT, Adolf Erman. Detailed older account, with much not in more recent books: domestic life, religion, magic, medicine, commerce, and whatever else needed for complete picture. Many illustrations. 597pp. 5⅜ × 8½.
22632-8 Pa. $8.95

HISTORIC COSTUME IN PICTURES, Braun & Schneider. Over 1,450 costumed figures shown, covering a wide variety of peoples: kings, emperors, nobles, priests, servants, soldiers, scholars, townsfolk, peasants, merchants, courtiers, cavaliers, and more. 256pp. 8⅜ × 11¼.
23150-X Pa. $9.95

THE NOTEBOOKS OF LEONARDO DA VINCI, edited by J. P. Richter. Extracts from manuscripts reveal great genius; on painting, sculpture, anatomy, sciences, geography, etc. Both Italian and English. 186 ms. pages reproduced, plus 500 additional drawings, including studies for *Last Supper*, *Sforza* monument, etc. 860pp. 7⅞ × 10¾. (Available in U.S. only) 22572-0, 22573-9 Pa., Two-vol. set $31.90

CATALOG OF DOVER BOOKS

THE ART NOUVEAU STYLE BOOK OF ALPHONSE MUCHA: All 72 Plates from "Documents Decoratifs" in Original Color, Alphonse Mucha. Rare copy-right-free design portfolio by high priest of Art Nouveau. Jewelry, wallpaper, stained glass, furniture, figure studies, plant and animal motifs, etc. Only complete one-volume edition. 80pp. 9⅜ × 12¼. 24044-4 Pa. $9.95

ANIMALS: 1,419 COPYRIGHT-FREE ILLUSTRATIONS OF MAMMALS, BIRDS, FISH, INSECTS, ETC., edited by Jim Harter. Clear wood engravings present, in extremely lifelike poses, over 1,000 species of animals. One of the most extensive pictorial sourcebooks of its kind. Captions. Index. 284pp. 9 × 12. 23766-4 Pa. $9.95

OBELISTS FLY HIGH, C. Daly King. Masterpiece of American detective fiction, long out of print, involves murder on a 1935 transcontinental flight—"a very thrilling story"—NY Times. Unabridged and unaltered republication of the edition published by William Collins Sons & Co. Ltd., London, 1935. 288pp. 5⅜ × 8½. (Available in U.S. only) 25036-9 Pa. $5.95

VICTORIAN AND EDWARDIAN FASHION: A Photographic Survey, Alison Gernsheim. First fashion history completely illustrated by contemporary photo-graphs. Full text plus 235 photos, 1840–1914, in which many celebrities appear. 240pp. 6½ × 9¼. 24205-6 Pa. $6.95

THE ART OF THE FRENCH ILLUSTRATED BOOK, 1700–1914, Gordon N. Ray. Over 630 superb book illustrations by Fragonard, Delacroix, Daumier, Doré, Grandville, Manet, Mucha, Steinlen, Toulouse-Lautrec and many others. Preface. Introduction. 633 halftones. Indices of artists, authors & titles, binders and provenances. Appendices. Bibliography. 608pp. 8⅜ × 11¼. 25086-5 Pa. $24.95

THE WONDERFUL WIZARD OF OZ, L. Frank Baum. Facsimile in full color of America's finest children's classic. 143 illustrations by W. W. Denslow. 267pp. 5⅜ × 8½. 20691-2 Pa. $7.95

FRONTIERS OF MODERN PHYSICS: New Perspectives on Cosmology, Rela-tivity, Black Holes and Extraterrestrial Intelligence, Tony Rothman, et al. For the intelligent layman. Subjects include: cosmological models of the universe; black holes; the neutrino; the search for extraterrestrial intelligence. Introduction. 46 black-and-white illustrations. 192pp. 5⅜ × 8½. 24587-X Pa. $7.95

THE FRIENDLY STARS, Martha Evans Martin & Donald Howard Menzel. Classic text marshalls the stars together in an engaging, non-technical survey, presenting them as sources of beauty in night sky. 23 illustrations. Foreword. 2 star charts. Index. 147pp. 5⅜ × 8½. 21099-5 Pa. $3.95

FADS AND FALLACIES IN THE NAME OF SCIENCE, Martin Gardner. Fair, witty appraisal of cranks, quacks, and quackeries of science and pseudoscience: hollow earth, Velikovsky, orgone energy, Dianetics, flying saucers, Bridey Murphy, food and medical fads, etc. Revised, expanded In the Name of Science. "A very able and even-tempered presentation."—The New Yorker. 363pp. 5⅜ × 8. 20394-8 Pa. $6.95

ANCIENT EGYPT: ITS CULTURE AND HISTORY, J. E Manchip White. From pre-dynastics through Ptolemies: society, history, political structure, religion, daily life, literature, cultural heritage. 48 plates. 217pp. 5⅜ × 8½. 22548-8 Pa. $5.95

SIR HARRY HOTSPUR OF HUMBLETHWAITE, Anthony Trollope. Incisive, unconventional psychological study of a conflict between a wealthy baronet, his idealistic daughter, and their scapegrace cousin. The 1870 novel in its first inexpensive edition in years. 250pp. 5⅜ × 8½. 24953-0 Pa. $5.95

LASERS AND HOLOGRAPHY, Winston E. Kock. Sound introduction to burgeoning field, expanded (1981) for second edition. Wave patterns, coherence, lasers, diffraction, zone plates, properties of holograms, recent advances. 84 illustrations. 160pp. 5⅜ × 8¼. (Except in United Kingdom) 24041-X Pa. $3.95

INTRODUCTION TO ARTIFICIAL INTELLIGENCE: SECOND, EN-LARGED EDITION, Philip C. Jackson, Jr. Comprehensive survey of artificial intelligence—the study of how machines (computers) can be made to act intelli-gently. Includes introductory and advanced material. Extensive notes updating the main text. 132 black-and-white illustrations. 512pp. 5⅜ × 8½. 24864-X Pa. $8.95

HISTORY OF INDIAN AND INDONESIAN ART, Ananda K. Coomaraswamy. Over 400 illustrations illuminate classic study of Indian art from earliest Harappa finds to early 20th century. Provides philosophical, religious and social insights. 304pp. 6⅜ × 9⅜. 25005-9 Pa. $9.95

THE GOLEM, Gustav Meyrink. Most famous supernatural novel in modern European literature, set in Ghetto of Old Prague around 1890. Compelling story of mystical experiences, strange transformations, profound terror. 13 black-and-white illustrations. 224pp. 5⅜ × 8½. (Available in U.S. only) 25025-3 Pa. $6.95

ARMADALE, Wilkie Collins. Third great mystery novel by the author of *The Woman in White* and *The Moonstone*. Original magazine version with 40 illustrations. 597pp. 5⅜ × 8½. 23429-0 Pa. $9.95

PICTORIAL ENCYCLOPEDIA OF HISTORIC ARCHITECTURAL PLANS, DETAILS AND ELEMENTS: With 1,880 Line Drawings of Arches, Domes, Doorways, Facades, Gables, Windows, etc., John Theodore Haneman. Sourcebook of inspiration for architects, designers, others. Bibliography. Captions. 141pp. 9 × 12. 24605-1 Pa. $7.95

BENCHLEY LOST AND FOUND, Robert Benchley. Finest humor from early 30's, about pet peeves, child psychologists, post office and others. Mostly unavailable elsewhere. 73 illustrations by Peter Arno and others. 183pp. 5⅜ × 8½. 22410-4 Pa. $4.95

ERTÉ GRAPHICS, Erté. Collection of striking color graphics: *Seasons, Alphabet, Numerals, Aces* and *Precious Stones*. 50 plates, including 4 on covers. 48pp. 9⅜ × 12¼. 23580-7 Pa. $6.95

THE JOURNAL OF HENRY D. THOREAU, edited by Bradford Torrey, F. H. Allen. Complete reprinting of 14 volumes, 1837–61, over two million words; the sourcebooks for *Walden*, etc. Definitive. All original sketches, plus 75 photographs. 1,804pp. 8½ × 12¼. 20312-3, 20313-1 Cloth., Two-vol. set $120.00

CASTLES: THEIR CONSTRUCTION AND HISTORY, Sidney Toy. Traces castle development from ancient roots. Nearly 200 photographs and drawings illustrate moats, keeps, baileys, many other features. Caernarvon, Dover Castles, Hadrian's Wall, Tower of London, dozens more. 256pp. 5⅜ × 8¼. 24898-4 Pa. $6.95

CATALOG OF DOVER BOOKS

AMERICAN CLIPPER SHIPS: 1833–1858, Octavius T. Howe & Frederick C. Matthews. Fully-illustrated, encyclopedic review of 352 clipper ships from the period of America's greatest maritime supremacy. Introduction. 109 halftones. 5 black-and-white line illustrations. Index. Total of 928pp. 5⅜ × 8½.
25115-2, 25116-0 Pa., Two-vol. set $17.90

TOWARDS A NEW ARCHITECTURE, Le Corbusier. Pioneering manifesto by great architect, near legendary founder of "International School." Technical and aesthetic theories, views on industry, economics, relation of form to function, "mass-production spirit," much more. Profusely illustrated. Unabridged translation of 13th French edition. Introduction by Frederick Etchells. 320pp. 6⅛ × 9¼. (Available in U.S. only)
25023-7 Pa. $8.95

THE BOOK OF KELLS, edited by Blanche Cirker. Inexpensive collection of 32 full-color, full-page plates from the greatest illuminated manuscript of the Middle Ages, painstakingly reproduced from rare facsimile edition. Publisher's Note. Captions. 32pp. 9⅜ × 12¼.
24345-1 Pa. $4.95

BEST SCIENCE FICTION STORIES OF H. G. WELLS, H. G. Wells. Full novel *The Invisible Man*, plus 17 short stories: "The Crystal Egg," "Aepyornis Island," "The Strange Orchid," etc. 303pp. 5⅜ × 8½. (Available in U.S. only)
21531-8 Pa. $6.95

AMERICAN SAILING SHIPS: Their Plans and History, Charles G. Davis. Photos, construction details of schooners, frigates, clippers, other sailcraft of 18th to early 20th centuries—plus entertaining discourse on design, rigging, nautical lore, much more. 137 black-and-white illustrations. 240pp. 6⅛ × 9¼.
24658-2 Pa. $6.95

ENTERTAINING MATHEMATICAL PUZZLES, Martin Gardner. Selection of author's favorite conundrums involving arithmetic, money, speed, etc., with lively commentary. Complete solutions. 112pp. 5⅜ × 8½.
25211-6 Pa. $2.95

THE WILL TO BELIEVE, HUMAN IMMORTALITY, William James. Two books bound together. Effect of irrational on logical, and arguments for human immortality. 402pp. 5⅜ × 8½.
20291-7 Pa. $7.95

THE HAUNTED MONASTERY and THE CHINESE MAZE MURDERS, Robert Van Gulik. 2 full novels by Van Gulik continue adventures of Judge Dee and his companions. An evil Taoist monastery, seemingly supernatural events; overgrown topiary maze that hides strange crimes. Set in 7th-century China. 27 illustrations. 328pp. 5⅜ × 8½.
23502-5 Pa. $6.95

CELEBRATED CASES OF JUDGE DEE (DEE GOONG AN), translated by Robert Van Gulik. Authentic 18th-century Chinese detective novel; Dee and associates solve three interlocked cases. Led to Van Gulik's own stories with same characters. Extensive introduction. 9 illustrations. 237pp. 5⅜ × 8½.
23337-5 Pa. $4.95

Prices subject to change without notice.
Available at your book dealer or write for free catalog to Dept. GI, Dover Publications, Inc., 31 East 2nd St., Mineola, N.Y. 11501. Dover publishes more than 175 books each year on science, elementary and advanced mathematics, biology, music, art, literary history, social sciences and other areas.